Praise

SP

"[Mr. Quammen] is not just among our best science writers but among our best writers, period. . . . That he hasn't won a nonfiction National Book Award or Pulitzer Prize is an embarrassment. . . . A patient explainer and a winning observer. His gallows humor is superb."

—Dwight Garner, *New York Times*

"A frightening and fascinating masterpiece."

—Walter Isaacson, author of *Steve Jobs*

"A literary crescendo." —April Dembosky, *Financial Times*

"Science writing as detective story at its best."

—Jennifer Ouellette, *Scientific American*'s
Cocktail Party Physics blog

"An adventure-filled page-turner . . . told from the front lines of pandemic prevention." —Lizzie Wade, *Wired*

"Masterful." —Florence Williams, *New York Review of Books*

"David Quammen [is] one of that rare breed of science journalists who blend exploration with a talent for synthesis and storytelling."

—Nathan Wolfe, *Nature*

"Quammen's more teacher than Jeremiah. So he calms when he can; but he's blunt when he must be."
—Jeffrey Burke, *Bloomberg*

"David Quammen might be my favorite living science writer: amiable, erudite, understated, incredibly funny, profoundly humane."
—Kathryn Schulz, *New York* magazine

"[*Spillover*] bucks the trend of one-hypothesis, cherry-picked, all-too-neat, over TED-dy fluff on the market."
—Ed Yong, creator of *National Geographic*'s
Not Exactly Rocket Science blog

"Not your typical sensationalist scare-a-thon. . . . [*Spillover*] hybridizes the rigor of an investigative with the suspense of a mystery novel, making for a page-turner that will have readers pondering the answers before the questions are even raised. . . . A remarkable achievement."
—*Science* magazine

ALSO BY DAVID QUAMMEN

NONFICTION

The Reluctant Mr. Darwin

Monster of God

The Song of the Dodo

Spillover

Ebola

ESSAYS

Natural Acts

The Boilerplate Rhino

Wild Thoughts from Wild Places

The Flight of the Iguana

FICTION

Blood Line

The Soul of Viktor Tronko

The Zolta Configuration

To Walk the Line

EDITED

On the Origin of Species by Charles Darwin: The Illustrated Edition

The Best American Science and Nature Writing 2000, with Burkhard Bilger

THE CHIMP AND THE RIVER

How AIDS Emerged
from an African Forest

DAVID QUAMMEN

W. W. NORTON & COMPANY
NEW YORK · LONDON

Extracted from *Spillover: Animal Infections and the Next Human Pandemic*, updated and with additional material.

Manufacturing by Courier Westford
Book design by Iris Weinstein
Production manager: Louise Parasmo

Library of Congress Cataloging-in-Publication Data

Quammen, David, 1948-, author.
The chimp and the river : how AIDS emerged from an African forest / David Quammen.
p. ; cm.
"Extracted from Spillover: Animal Infections and the Next Human Pandemic, updated and with additional material."—supplied by publisher.
Includes bibliographical references and index.
ISBN 978-0-393-35084-5 (pbk.)
I. Quammen, David, 1948– . Spillover : animal infections and the next human pandemic. Based on (work): II. Title.
[DNLM: 1. Acquired Immunodeficiency Syndrome—etiology—Popular Works. 2. HIV—pathogenicity—Popular Works. 3. Zoonoses—Popular Works. WC 503.3]
QR201.A37
614.5'99392—dc23

2014042399

W. W. Norton & Company, Inc.
500 Fifth Avenue, New York, N.Y. 10110
www.wwnorton.com

W. W. Norton & Company Ltd.
Castle House, 75/76 Wells Street, London W1T 3QT

1 2 3 4 5 6 7 8 9 0

To the men and women who have suffered this disease and those who continue to suffer it

CONTENTS

INTRODUCTION

AIDS has been written about many times, from many angles, but the story as told here is drastically different from any version you are likely to have read, heard about, or otherwise absorbed. The main difference is that this is an account of the ultimate source of the pandemic. It's essentially an ecological narrative rather than a medical one—by which I mean that it turns on the interaction of three kinds of creature: chimpanzee, human, and virus. I have tried to describe how, when, and where the virus in question (HIV-1 group M, the pandemic strain), which was previously unknown among people, got its start in the human population. There was a single point in space and time. There was a fateful event. The world became cognizant of AIDS in the early 1980s, of course, but that wasn't the beginning—far from it. The real moment of origin occurred decades earlier and thousands of miles away. It's now possible to say, based on persuasive research in molecular genetics, as published in scientific journals but little noticed by the public, that the virus passed from a single chimpanzee into a single person, during a presumably bloody encounter, in the southeastern corner of Cameroon, near a minor tributary of the Congo River, around 1908, give or take a margin of error. Key components of that research were led by Beatrice Hahn, then of the University of Alabama at Birmingham, and Michael Worobey, at the University of Arizona, both of whom feature in this little book.

In *The Chimp and the River* I trace backward to that moment of origin, reconstruct it, and then follow its consequences forward again, along the pathways of social history, epidemiology, and dolorous accident, to the point when AIDS "suddenly" emerged as a global human disaster.

AIDS is horrible and unique, but it's also part of a larger pattern. Everything comes from somewhere, and strange new infectious diseases, emerging abruptly among humans, come mostly from nonhuman animals. The disease might be caused by a virus, or a bacterium, or a protozoan, or some other form of dangerous bug. That bug might live inconspicuously in a kind of rodent, or a bat, or a bird, or a monkey, or an ape. Crossing by happenstance from its animal hideaway into its first human victim, it might find surprisingly hospitable conditions; it might replicate aggressively and abundantly; it might cause illness, even death; and in the meantime, it might pass onward from its first human victim into others. There's a fancy term for this phenomenon, used by scientists who study infectious diseases from an ecological perspective: *zoonosis*. A zoonosis is an animal infection transmissible to humans. The animal hideaway is known as a reservoir host. The event of transmission, from one species into another, is called spillover.

That bit of terminology gave me the title for my 2012 book, *Spillover: Animal Infections and the Next Human Pandemic*, in which *The Chimp and the River* first appeared as a long chapter. I revised that material here just enough to make it a self-contained narrative with its own wider meaning and its own dire implication. Its wider meaning is that pandemic AIDS resulted from our dealings with the natural world, as well as from our dealings with one another. Its dire implication is that, having happened once catastrophically, and at least eleven other times (as you'll learn from this story) less consequentially, the spillover of an HIV-like virus from nonhuman primates to humans is not a highly improbable event. It could certainly happen again.

Spillover is a book about zoonotic diseases and their increasing importance in our modern world—a world where 7 billion humans interact ever more frequently, ever more disruptively, with the wild animals that live in wild places. Roughly 60 percent of the infectious diseases known among humans are zoonotic in the strict sense, coming to us continually or within recent history from animals; most if not all of the other 40 percent, including human afflictions such as measles and smallpox, can be considered zoonotic in a broader sense, given that we are a relatively young species and our tormenting bugs have their own histories that precede us. Everything, as I've said, and will say again, comes from somewhere.

My lengthy treatment of this subject, in *Spillover*, focuses on a selection of nasty diseases, some new and some old, some infamous and some obscure, and on the viruses and other microbes that cause them: Ebola in Africa, Nipah in Bangladesh and Malaysia, SARS as it came out of China, Lyme disease in the suburbs of New England, Hendra amid the horse cultures of Australia, zoonotic malaria in Borneo, influenza everywhere, and a few others. SARS killed about eight hundred people during its frightening emergence back in 2003, but then it was controlled by good science and rigorous public health measures, and the SARS virus has not re-emerged since. Ebola virus disease is gruesome and highly lethal but its geographical reach and its total of human victims, so far, have been relatively small (although, at the time of this writing, the reach of Ebola has increased alarmingly during the 2014 outbreak in West Africa; no one yet knows where that will end). Lyme disease is punishingly familiar to many people but it doesn't usually kill them. Nipah has re-emerged repeatedly in Bangladesh, usually when the seasonal harvest of date-palm sap has brought sap harvesters and their customers into unhealthy contact with the excretions of bats. Hendra can be terrifying if you're an Australian who works with horses, but it has killed fewer than ten people since its first known occurrence in 1994. Given the limited scope and peculiarities of such problems, why should people around the rest of the world concern themselves with the subject of zoonotic disease?

It's a fair question but there are good answers. Some of those answers are intricate and speculative. Others are objective and blunt. The bluntest is this: AIDS.

THE CHIMP AND THE RIVER

1

There are multiple beginnings to what we think we know about the AIDS pandemic, most of which don't even address the subject of its ultimate origin in a single chimpanzee.

For instance: In autumn of 1980, a young immunologist named Michael Gottlieb, an assistant professor at the UCLA Medical Center, began noticing a strange pattern of infections among certain male patients. The patients, eventually five of them, were all active homosexuals and all suffering from pneumonia caused by a usually harmless fungus then known as *Pneumocystis carinii.* (Nowadays, after a name change, it's *Pneumocystis jurovecii.*) The stuff is ubiquitous; it floats around everywhere. Their immune systems should have been able to clear it. But their immune systems evidently weren't working, and this fungus filled their lungs. Each man also had another sort of fungal infection—oral candidiasis, meaning a mouthful of slimy *Candida* yeast, more often seen in newborn babies, diabetics, and people with compromised immune systems than in healthy adults. Blood tests, done on several of the patients, showed dramatic depletions of certain lymphocytes (white blood cells) that are crucial in regulating immune responses. Specifically, it was thymus-dependent lymphocytes (T cells, for short) that were "profoundly depressed" in number. Although Gottlieb noted some other symptoms, those three stood out: *Pneumocystis* pneumonia, oral candidiasis, dearth of T cells. In mid-May of 1981, he and a colleague wrote a brief paper, with cooperation from other Los Angeles doctors, describing their observations. They didn't speculate about causes. They just saw the pattern as a befuddling, ominous trend and felt they should publish quickly. An editor at *The New England Journal of Medicine* was interested but his lead time would be at least three months.

So Gottlieb turned to the streamlined newsletter of the Centers for Disease Control (as that agency was then known) in Atlanta. The CDC issues *Morbidity and Mortality Weekly Report* to deliver breaking news of disease events in a timely way. Gottlieb's barebones text, less than two pages long, appeared in *MMWR* on June 5, 1981, under the dry title "*Pneumocystis* Pneumonia—Los Angeles." It was the first published medical alert about a syndrome that didn't yet have a name.

The second alert came a month later, again in the CDC newsletter. While Gottlieb noticed *Pneumocystis* pneumonia and candidiasis, a New York dermatologist named Alvin E. Friedman-Kien spotted a parallel trend involving a different disease: Kaposi's sarcoma. A rare form of cancer, not usually too aggressive, Kaposi's sarcoma was known primarily as an affliction of middle-aged Mediterranean males—the sort of fellows you'd expect to find in an Athens café, drinking coffee and playing dominoes. This cancer often showed itself as purplish nodules in the skin. Within less than three years, Friedman-Kien and his network of colleagues saw twenty-six cases of Kaposi's sarcoma in youngish homosexual men. Some of those patients also had *Pneumocystis* pneumonia. Eight of them died. Hmm. *Morbidity and Mortality Weekly Report* carried Friedman-Kien's communication on July 3, 1981.

Kaposi's sarcoma also figured prominently in a set of clinical observations made in Miami around the same time. The symptoms among this group of patients were similar; the cultural profile was different. These sick people, twenty of them, hospitalized between early 1980 and June 1982, were all Haitian immigrants. Most had arrived in the United States recently. By their own testimony during medical interviews they were all heterosexuals, with no history of homosexual activity. But their cluster of ailments resembled what Gottlieb had seen among gay men in Los Angeles and Friedman-Kien among gay men in New York: *Pneumocystis* pneumonia, candidiasis in the throat, plus other unusual infections, irregularities in lymphocyte counts, and aggressive Kaposi's sarcoma. Ten of the Haitians died. The team of doctors who published these observations saw a "syndrome" that seemed "strikingly similar to the syndrome of immunodeficiency described recently among American homosexuals." The early connection to Haitian heterosexuals would later come to seem like a false lead and be largely ignored in discussions of AIDS. It was hard to confirm, given the limits of interview data, and harder still to construe. Calling attention to it even came to seem politically incorrect. Then, later still, its real significance would emerge from work at the level of molecular genetics.

Another perceived starting point was Gaëtan Dugas, the young Canadian flight attendant who became notorious as "Patient Zero." You've heard of him, probably, if you've heard much of anything about the dawning of AIDS. Dugas has been written about as the man who "carried the virus out of Africa and introduced it into the Western gay community." He wasn't. But he seems to have played an oversized and culpably heedless role as a transmitter during the 1970s and early 1980s. As a flight steward, with almost cost-free privileges of personal travel, he flew often between major cities in North America, joining in sybaritic play where he landed, notching up conquests, living the high life of a sexually voracious gay man at the height of the bathhouse era. He was handsome, sandy-haired, vain but charming, even "gorgeous" in some eyes. According to Randy Shilts, author of *And the Band Played On* (which includes much heroic research and a bit of questionably reliable reimagining), Dugas himself reckoned that in the decade since becoming actively gay he had had at least twenty-five hundred sexual partners. Dugas paid a price for his appetite and his daring. He developed Kaposi's sarcoma, underwent chemotherapy for that, suffered from *Pneumocystis* pneumonia and other AIDS-related infections, and died of kidney failure at age thirty-one. During the brief stretch of years between his Kaposi's diagnosis and his final invalidism, Gaëtan Dugas didn't slow down. But he seems to have tipped, in his lonely despair, from hedonism to malice; he would have sex with a new acquaintance at the Eighth-and-Howard bathhouse in San Francisco, then turn up the lights—so Randy Shilts claimed—display his lesions, and say: "I've got gay cancer. I'm going to die and so are you."

In the same month as Dugas's death, March 1984, a team of epidemiologists from the CDC published a landmark study of the role of sexual contact in linking cases of what by then was called AIDS. The world had a label now but not an explanation. "Although the cause of AIDS is unknown," wrote the CDC team, whose lead author was David M. Auerbach, "it may be caused by an infectious agent that is transmissible from person to person in a manner analogous to hepatitis B infection." Hepatitis B is a blood-borne virus. It moves primarily by sexual contact, intravenous drug use with shared needles, or transfusion of blood products carrying the virus as a contaminant. It seemed like a template for understanding what otherwise was still a bewildering convergence of symptoms. "The existence of a cluster of AIDS cases linked by homosexual contact is consistent with an infectious-agent hypothesis," the CDC group added. Not a toxic chemical, not an accident of genetics, but some kind of bug, is what they meant.

Auerbach and his colleagues had gathered information from nineteen AIDS cases in southern California, interviewing each patient or, if he was dead, his close companions. They spoke with another twenty-one patients in New York and other American cities, and from their forty case histories they created a graphic figure of forty interconnected disks, like a Tinkertoy structure, showing who was linked sexually with whom. The patients' identities were coded by location and number, such as "SF 1," "LA 6," and "NY 19." At the center of the network, connected directly to eight disks and indirectly to all the rest, was a disk labeled "0." Although the researchers didn't name him, that patient was Gaëtan Dugas. Randy Shilts later transformed the somewhat bland "Patient 0," as mentioned in this paper, to the more resonant "Patient Zero" of his book. But what the word "Zero" belies, what the number "0" ignores, and what the central position of that one disk within the figure fails to acknowledge, is that Gaëtan Dugas didn't conceive the AIDS virus himself. Everything comes from somewhere, and he got it from someone else. Dugas himself was infected by some other human, presumably during a sexual encounter—and not in Africa, not in Haiti, somewhere closer to home. That was possible because, as evidence now shows, HIV had already arrived in North America when Gaëtan Dugas was a virginal adolescent.

It had also arrived in Europe, though on that continent it hadn't yet gone far. A Danish doctor named Grethe Rask, who had been working in Africa, departed in 1977 from what was then Zaire and returned to Copenhagen for treatment of a condition that had been dragging her downward for several years. During her time in Zaire, Rask first ran a small hospital in a remote town in the north, then served as chief surgeon at a large Red Cross facility in the capital, Kinshasa. Somewhere along the way, possibly during a surgical procedure done without adequate protective supplies (such as rubber gloves), she became infected with something for which no one at the time had a description or a name. She felt ill and fatigued. Drained by persistent diarrhea, she lost weight. Her lymph nodes swelled and stayed swollen. She told a friend: "I'd better go home to die." Back in Denmark, tests revealed a shortage of T cells. Her breath came with such difficulty that she depended on bottled oxygen. She struggled against staph infections. *Candida* fungus glazed her mouth. By the time Grethe Rask died, on December 12, 1977, her lungs were clogged with *Pneumocystis carinii*, and that seems to have been what killed her.

It shouldn't have, according to standard medical wisdom. *Pneumocystis* pneumonia wasn't normally a fatal condition. There had to be a broader explanation, and there was. Nine years later, a sample of Rask's blood serum tested positive for HIV.

All these unfortunate people—Grethe Rask, Gaëtan Dugas, the five men in Gottlieb's report from Los Angeles, the Kaposi's sarcoma patients known to Friedman-Kien, the Haitians in Miami, the cluster of thirty-nine (besides Dugas) identified in David Auerbach's study—were among the earliest recognized cases of what has retrospectively been identified as AIDS. But they weren't among the first victims. Not even close. Instead they represent midpoints in the course of the pandemic, marking the stage at which a slowly building, almost unnoticeable phenomenon suddenly rose to a crescendo. The real beginning of AIDS lay elsewhere, and more decades passed while a few scientists worked to discover it.

2

In the early years after its detection, the new illness was a shifting shape that carried several different names and acronyms. GRID was one, standing for Gay-Related Immune Deficiency. That proved too restricted as heterosexual patients began to turn up: needle-sharing addicts, hemophiliacs, other unlucky straights. Some doctors preferred ACIDS, for Acquired Community Immune Deficiency Syndrome. The word "community" was meant to signal that people acquired it *out there*, not in hospitals. A more precise if clumsier formulation, favored briefly by the CDC's *Morbidity and Mortality Weekly Report,* was "Kaposi's sarcoma and opportunistic infections in previously healthy persons," which didn't abbreviate neatly. KSOIPHP lacked punch. By September 1982, *MMWR* had switched its terminology to Acquired Immune Deficiency Syndrome (AIDS), and the rest of the world followed.

Naming the syndrome was the least of the early challenges. More urgent was to identify its cause. No one knew, back when those reports from Gottlieb and Friedman-Kien began capturing attention, what sort of pathogen caused this combination of puzzling, lethal symptoms—nor even if there *was* a single pathogen. The virus idea arose, after other mistaken hypotheses, as a plausible guess.

One scientist who made the guess was Luc Montagnier, then a little-known molecular biologist at the Institut Pasteur in Paris. Montagnier's research focused mainly on cancer-causing viruses, especially the group known as retroviruses, some of which cause tumors in birds and mammals. Retroviruses are fiendish things, even more devious and persistent than the average virus. They take their name from the capacity to move backward (retro) against the usual expectations of how a creature translates its genes into working proteins. Instead of using RNA as a template for translating DNA into proteins—the usual route by which genetic information becomes living reality—a retrovirus converts its RNA into DNA within a host cell; its viral DNA then penetrates the cell nucleus and gets itself integrated into the genome of the host cell, thereby guaranteeing replication of the virus whenever the host cell reproduces itself. Luc Montagnier had studied these things in animals—chickens, mice, primates—and wondered about the possibility of finding them in human tumors too. Another disquieting possibility about retroviruses was that the new disease showing up in America and Europe, AIDS, might be caused by one.

There was still no solid proof that it was caused by a virus of any sort. But three kinds of evidence pointed that way, and Montagnier recalled them in his memoir, a book titled *Virus*. First, the incidence of AIDS among homosexuals linked by sexual interactions suggested that this was an infectious disease. Second, the incidence among intravenous drug users suggested a blood-borne infectious agent. Third, the cases among hemophiliacs implied a blood-borne agent that escaped detection in processed blood products such as clotting factor. So: It was contagious, blood-borne, infinitesimal.

"AIDS could not be caused by a conventional bacterium, a fungus, or protozoan," Montagnier wrote, "since these kinds of germs are blocked by the filters through which the blood products necessary to the survival of hemophiliacs are passed. That left only a smaller organism: the agent responsible for AIDS thus could only be a virus."

Other evidence hinted that, among all viral possibilities, it might be a retrovirus. This was new ground, but then so was AIDS. The only known human retrovirus as of early 1981 was something called human T-cell leukemia virus (HTLV), recently discovered under the leadership of a smart, outgoing, highly regarded, and highly ambitious researcher named Robert Gallo, whose Laboratory of Tumor Cell Biology was part of the National Cancer Institute in Bethesda, Maryland. HTLV, as its name implies, attacks T cells and can turn them cancerous. T cells are one of the three major types of lymphocyte of the immune system. (Later the acronym HTLV was recast to mean human T-lymphotropic virus, which is slightly more accurate.) A related retrovirus, feline leukemia virus, causes immunodeficiency in cats. So a suspicion arose among cancer-virus researchers that the AIDS agent, destroying human immune systems by attacking their lymphocytes (in particular, a subcategory of T cells known as T-helper cells), might likewise be a retrovirus. Montagnier's group began looking for it.

Gallo's lab did too. And those two weren't alone. Other scientists at other laboratories around the world recognized that finding the cause of AIDS was the hottest, the most urgent, and potentially the most rewarding quest in medical research. By late spring of 1983, three teams working independently had each isolated a candidate virus, and in the May 20 issue of *Science*, two of those teams published announcements. Montagnier's group in Paris, screening cells from a thirty-three-year-old homosexual man suffering from lymphadenopathy (swollen lymph nodes), had found a new retrovirus, which they called LAV (for lymphadenopathy virus). Gallo's group came up with a new virus also, one that Gallo took for a close relative of the human T-cell leukemia viruses (by now there was a second, called HTLV-II, and the first had become HTLV-I) that he and his people had discovered. He called this newest bug HTLV-III, nesting it proprietarily into his menagerie. The French LAV and the Gallo HTLVs had at least one thing in common: They were indeed retroviruses. But within that family exists some rich and important diversity. An editorial in the same issue of *Science* trumpeted the Gallo and Montagnier papers with a misleading headline: HUMAN T-CELL LEUKEMIA VIRUS LINKED TO AIDS, despite the fact that Montagnier's LAV was *not* a human T-cell leukemia virus. Woops, mistaken identity. Montagnier knew better, but his *Science* paper seemed to blur the distinction, and the editorial occluded it entirely.

Then again, neither was Gallo's "HTLV-III" an HTLV, once it was clearly seen and correctly classified. It turned out to be something nearly identical to Montagnier's LAV, of which Montagnier had given him a frozen sample. Montagnier had personally delivered that sample, carrying it on dry ice during a visit to Bethesda.

Confusion was thus sown early—confusion about what exactly had been discovered, who had discovered it, and when. That confusion, irrigated with competitive zeal, fertilized with accusation and denial, would grow rife for decades. There would be lawsuits. There would be fights over royalties from the patent on an AIDS blood-screening test that derived from virus grown in Gallo's lab but traceable to Montagnier's original isolate. (Contamination from one experiment to another, or from one batch of samples to another, is a familiar problem in lab work with viruses.) It wasn't a petty squabble. It was a big squabble, in which pettiness played no small part. What was ultimately at stake, besides money and ego and national pride, was not just advancing or retarding research toward an AIDS cure or vaccine but also the Nobel Prize in medicine, which eventually went to Luc Montagnier and his chief collaborator, Françoise Barré-Sinoussi.

Meanwhile the third team of researchers, led quietly by a man named Jay A. Levy in his lab at the University of California School of Medicine, in San Francisco, also found a candidate virus in 1983 but didn't publish until more than a year afterward. By summer of 1984, Levy noted, AIDS had affected "more than 4000 individuals in the world; in San Francisco, over 600 cases have been reported." Those numbers sounded alarmingly high at the time, though in retrospect, compared with 36 million deaths, they seem poignantly low. Levy's discovery was also a retrovirus. His group detected it in twenty-two AIDS patients and grew more than a half dozen isolates. Because the bug was an AIDS-associated retrovirus, Levy called it ARV. He suspected, correctly, that his ARV and Montagnier's LAV were simply variant samples of the same evolving virus. They were very similar but not *too* similar. "Our data cannot reflect a contamination of our cultures with LAV," he wrote, "since the original French isolate was never received in our laboratory." Harmless as that may sound, it was an implicit jab at Robert Gallo.

The details of this story, the near-simultaneous triple discovery and its aftermath, are intricate and contentious and seamy and technical, like a ratatouille of molecular biology and personal politics left out in the sun to ferment. They lead far afield from the subject of zoonotic disease. For our purposes here, the essential point is that a virus discovered in the early 1980s, in three different places under three different names, became persuasively implicated as the causal agent of AIDS. A distinguished committee of retrovirologists settled the naming issue in 1986. They decreed that the virus would be called HIV.

3

It's quite appropriate and more than coincidental, given the zoonotic dimension of this story, that the next phase began with a veterinarian. Max Essex studied retroviruses in monkeys and cats.

Dr. Myron (Max) Essex, DVM, PhD, was not your ordinary small-animal vet. He was a professor and a research scientist in the Department of Cancer Biology at the Harvard School of Public Health. He had worked on feline leukemia virus (FeLV), among other things, and cancer-causing viruses formed the broad frame of his interests. Having seen the effects of FeLV in wrecking the immune systems of cats, he suspected as early as 1982, along with Gallo and Montagnier, that the new human immune deficiency syndrome might be caused by a retrovirus.

Then something strange came to his notice, by way of a grad student named Phyllis Kanki. She was a veterinarian like him, but now working on a doctorate there at the School of Public Health. Kanki grew up in Chicago, spent her adolescent summers doing zoo work, then studied biology and chemistry on the way toward veterinary medicine and comparative pathology. During the summer of 1980, while still amid her DVM studies, she worked at the New England Regional Primate Research

Center, which was part of Harvard but located out in Southborough, Massachusetts. There she saw a weird problem among the center's captive Asian macaques—some of them were dying of a mysterious immune dysfunction. A macaque is a kind of monkey, and several of the Asian species, such as the rhesus macaque and the Formosan rock macaque, are highly valued as laboratory animals for medical research. At the primate center, which held almost eight hundred animals of the macaque genus, Formosan rock macaques in particular seemed to be suffering this immune-system failure. Their T-helper lymphocyte counts were way down. They wasted away from diarrhea or succumbed to opportunistic infections, including *Pneumocystis carinii.* It sounded too much like AIDS. Kanki later brought this to the attention of Essex, her thesis adviser, and together with colleagues from Southborough, they started to look for what was killing those monkeys. Based on their knowledge of FeLV and other factors, they wondered whether it might be a retrovirus infection.

Taking blood samples from macaques, they did find a new retrovirus, and saw that it was closely related to the AIDS virus. Because this was 1985, they used Gallo's slightly misleading label (HTLV-III) for what would soon be renamed HIV. Their monkey virus would be renamed too and become, by analogy, simian immunodeficiency virus: SIV. The group published a pair of papers in *Science*, which had grown hungry for AIDS breakthroughs. This discovery, they wrote, could help illuminate the pathology of the disease, maybe even advance efforts to develop a vaccine, by providing an animal model for research. Only a single sentence at the end of one of the papers, a modest but pertinent comment dropped in like an afterthought, noted that SIV might also be a clue toward the *origin* of HIV.

It was. Phyllis Kanki performed the lab analysis of samples

from the captive macaques and then made it her business to wonder whether the same virus might exist in the wild. Kanki and Essex looked at other Asian macaques, testing blood samples from wild-caught animals. They found no trace of SIV. They tested still other kinds of wild Asian monkey. Again, no SIV. This led them to surmise that the macaques at Southborough had picked up their SIV in captivity by exposure to animals of another species. It was a reasonable guess, given that the primate center at one point had a monkey playpen in its lobby, where Asian and African infant monkeys were sometimes allowed to mingle. But then which kind of African monkey was the reservoir? Where exactly had the virus come from? And how might it be related to the emergence of HIV?

"In 1985, the highest rates of HIV were reported in the U.S. and Europe," Essex and Kanki wrote later, "but disturbing reports from central Africa indicated that high rates of HIV infection and of AIDS prevailed there, at least in some urban centers." The focus of suspicion was shifting: not Asia, not Europe, not the United States, but *Africa* might be the point of origin. Central Africa also harbored a rich fauna of nonhuman primates. So the Harvard group got hold of blood from some wild-caught African simians, including chimpanzees, baboons, and African green monkeys. None of the chimps or the baboons showed any sign of SIV infection. Some of the African green monkeys did. It was an epiphany. More than two dozen of the monkeys carried antibodies to SIV, and Kanki grew isolates of live virus from seven. That finding too went straight into *Science,* and the search continued. Kanki and Essex eventually screened thousands of African green monkeys, caught in various regions of sub-Saharan Africa or held captive in research centers around the world. Depending on the population, between 30 and 70 percent of those animals tested SIV-positive.

But the monkeys weren't sick. They didn't seem to be suffering from immunodeficiency. Unlike the Asian macaques, the African green monkeys "must have evolved mechanisms that kept a potentially lethal pathogen from causing disease," Essex and Kanki wrote. Maybe the virus had changed too. "Indeed, some SIV strains might also have evolved toward coexistence with their monkey hosts." The monkeys evolving toward greater resistance, the virus evolving toward lesser virulence—this sort of mutual adaptation would suggest that SIV had been in them a long time.

The new virus, SIV as found in African green monkeys, became the closest known relative of HIV. But it wasn't *that* close; many differences distinguished the two at the level of genetic coding. The resemblance, according to Essex and Kanki, was "not close enough to make it likely that SIV was an immediate precursor of HIV in people." More likely, those two viruses represented neighboring twigs on a single phylogenetic branch, separated by lots of evolutionary time and probably some extant intermediate forms. Where might the missing cousins be? "Perhaps, we thought, one could find such a virus—an intermediate between SIV and HIV—in human beings." They decided to look in West Africa.

With help from an international team of collaborators, Kanki and Essex gathered blood samples from Senegal and elsewhere. The samples arrived with coded labeling, for blind testing in the laboratory, so that Kanki herself didn't know their country of origin, nor even whether they derived from humans or monkeys. She screened them using tests for both SIV and HIV. Despite one possible misstep involving a lab contamination, her team found what they had thought they might: a virus intermediate between HIV and SIV. With the code unblinded, Kanki learned that the positive results came from Senegalese prostitutes. In retrospect it made sense. Prostitutes are at high risk for any sexually transmitted virus, including a new one recently spilled into humans. And the density of the rural human population in Senegal, where African green monkeys are native, makes monkey-human interactions (crop-raiding by monkeys, hunting by humans) relatively frequent.

Furthermore, the new bug from Senegalese prostitutes wasn't just halfway between HIV and SIV. It more closely resembled SIV strains from African green monkeys than it did the Montagnier-Gallo version of HIV. That was important but puzzling. Were there two distinct kinds of HIV?

Luc Montagnier now reenters the story. Having tussled with Gallo over the first HIV discovery, he converged more amicably with Essex and Kanki on this one. Using assay tools provided by the Harvard group, Montagnier and his colleagues screened the blood of a twenty-nine-year-old man from Guinea-Bissau, a tiny country, formerly a Portuguese colony, along the south border of Senegal. This man showed symptoms of AIDS (diarrhea, weight loss, swollen lymph nodes) but tested negative for HIV. He was hospitalized in Portugal, and his blood sample was hand-delivered to Montagnier by a visiting Portuguese biologist. In Montagnier's lab, the man's serum again tested negative for antibodies to HIV. But from a culture of his white blood cells Montagnier's group isolated a new human retrovirus, which looked very similar to what Essex and Kanki had found. In another patient, hospitalized in Paris but originally from Cape Verde, an island nation off the west coast of Senegal, the French team found more virus of the same type. Montagnier called the new thing LAV-2. Eventually, when all parties embraced the label HIV instead, it would be HIV-2. The original became HIV-1.

The paths of discovery may be sinuous, the labels may seem

many, and maybe you can't tell the players without a scorecard; but these details aren't trivial. The difference between HIV-2 and HIV-1 is the difference between a nasty little West African disease and a global pandemic.

4

During the late 1980s, as Phyllis Kanki and Max Essex and other scientists studied HIV-2, a flurry of uncertainty arose about its provenance. Some challenged the idea that it was closely related to (and recently derived from) a retrovirus that infects African monkeys. An alternative view was that such a retrovirus had been present in the human lineage for as long as—or longer than—we have existed as a species. Possibly it was already with us, a passenger riding the slow channels of evolution, when we diverged from our primate cousins. But that view left an unresolved conundrum: If the virus was an ancient parasite upon humans, unnoticed for millennia, how had it suddenly become so pathogenic?

Recent spillover seemed more likely. Still, the case *against* that idea got a boost in 1988, when a group of Japanese researchers sequenced the complete genome of SIV from an African green monkey. That is, they detected and assembled the linear message of nucleotide bases—represented by the letters A, T, C, and G—comprising that SIV's genetic inscription. The host animal came from Kenya. The nucleotide sequence of its retrovirus proved to be substantially different from the sequence for HIV-1, and different in roughly the same degree from HIV-2. So the monkey virus seemed no more closely related to the one human virus than to the other. That contradicted the notion that HIV-2 had lately emerged from an African green monkey. A commentary in the journal *Nature*, published to accompany the Japanese paper, celebrated this finding beneath a dogmatic headline: HUMAN AIDS VIRUS NOT FROM MONKEYS. But the headline was misleading to the point of falsity. *Not from monkeys*? Well, don't be so sure. It turned out that researchers were just looking at the wrong kind of monkey.

Confusion came from two sources. For starters, the label "African green monkey" is a little vague. It encompasses a diversity of forms, sometimes also known as savannah monkeys, that occupy adjacent geographical ranges spread across sub-Saharan Africa, from Senegal in the west to Ethiopia in the east and down into South Africa. At one time those forms were considered a "superspecies" under the name *Cercopithecus aethiops*. Nowadays, their differences having been more acutely gauged, they are classified as six distinct species within the genus *Chlorocebus*. The "African green monkey" sampled by the Japanese team, because it was "of Kenyan origin," probably belonged to the species *Chlorocebus pygerythrus*. The species native to Senegal, on the other hand, is *Chlorocebus sabaeus*. Now that you've seen those two names you can forget them. The difference between one African green monkey and another is not what accounts for the genetic disjunction between SIV and HIV-2.

The trail backward from HIV-2 led to another monkey entirely: the sooty mangabey. This is not one of the six *Chlorocebus* species, not even close. It belongs to a different genus.

The sooty mangabey (*Cercocebus atys*) is a smoky-gray creature with a dark face and hands, white eyebrows, and flaring white muttonchops, not nearly so decorative as many monkeys on the continent but arresting in its way, like an elderly chimney sweep of dapper tonsorial habits. It lives in coastal West Africa, from Senegal to Ghana, favoring swamps and palm forests, where it eats fruit, nuts, seeds, leaves, shoots, and roots—an eclectic vegetarian—and spends much of its time on the ground, moving quadrupedally in search of fallen tidbits. Sometimes it ventures out of the bottomlands to raid farms and rice paddies. The sooty mangabey is hard to hunt within the swampy forests but, because of its terrestrial foraging habits and its taste for crops, easy to trap. Local people treat it as an annoying but edible sort of vermin. Sometimes also, if they're not too hungry, they adopt an orphan juvenile as a pet.

What brought the sooty mangabey to the attention of AIDS researchers was coincidence and an experiment on leprosy. It was an instance of the old scientific verity that sometimes you find much more than you're looking for.

Back in September 1979, scientists at a primate research center in New Iberia, Louisiana, south of Lafayette, had noticed a leprosy-like infection in one of their captive monkeys. This seemed odd, because leprosy is a human disease caused by a bacterium (*Mycobacterium leprae*) not known to be transmissible from people to other primates. But here was a leprous monkey. The animal in question, a sooty mangabey, female, about five years old, had been imported from West Africa. The researchers called her Louise. Apart from her skin condition, Louise was healthy. She hadn't, so far as the records showed, yet been subjected to any experimental infection. They were using her in a study of diet and cholesterol. The New Iberia facility didn't happen to work on leprosy infections, so once Louise's condition had been recognized she was transferred to a place, also in Louisiana, that did: the Delta Regional Primate Research Center, north of Lake Pontchartrain. The researchers at Delta were glad to get her, for one very practical reason. If Louise had acquired her leprosy naturally, then (contrary to previous suppositions) the disease might be transmissible in populations of sooty mangabey. And if that were true, then the sooty mangabey could prove valuable as an experimental model for studies of human leprosy. This is how human medical research works: at the expense of other creatures.

So the Delta team injected some infectious material from Louise into another sooty mangabey. This one was a male. Unlike Louise, he is nameless in the scientific record, remembered only by a code: A022. He became the first in a chain of experimentally infected monkeys that turned out to carry more than leprosy. The scientists at Delta had no idea, not at first, that A022 was SIV-positive.

The leprosy from Louise took hold easily in A022, which was notable, given that earlier attempts to infect monkeys with human leprosy had failed. Was this strain of *Mycobacterium leprae* a peculiarly monkey-adapted variant? If so, might it succeed in rhesus macaques too? That would be convenient for experimental purposes, because rhesus macaques were cheaper and more available, in the medical-research chain of supply, than sooty mangabeys. So the Delta team injected four rhesus macaques with infectious gunk from A022. All four developed leprosy. For three of the four, that proved to be the least of their troubles. The unlucky three also developed simian AIDS. Suffering chronic diarrhea and weight loss, they wasted away and died.

Screening for virus, the researchers found SIV. How had their three macaques become SIV-positive? Evidently by way of the leprous inoculum from the sooty mangabey, A022. Was he unique? No. Tests of other sooty mangabeys at Delta revealed that the virus was quite prevalent among them. Other investigators soon found it too, not just among captive sooty mangabeys but also in the wild. Yet the sooty mangabeys (native to Africa), unlike the Asian macaques, showed no symptoms of simian AIDS. They were infected but healthy, which suggested that the virus had a long history in their kind. The same virus made the macaques sick, presumably because it was new to them.

The roster of simian immunodeficiency viruses was growing more crowded and complex. Now there were three known variants: one from African green monkeys, one from rhesus macaques (which they probably acquired in captivity), and one from sooty mangabeys. Needing a way to identify and distinguish them, someone hit upon the expedient of adding tiny subscripts to the acronym. Simian immunodeficiency virus as found in sooty mangabeys became SIV_{sm}. The other two were labeled SIV_{agm} (for African green monkeys) and SIV_{mac} (for Asian macaques). This little convention may seem esoteric, not to mention hard on the eyes, but it will be essential and luminous when I discuss the fateful significance of a variant that came to be known as SIV_{cpz}.

For now it's enough to note the upshot of the leprosy experiment in Louisiana. One scientist from the Delta team, a woman named Michael Anne Murphey-Corb, collaborated with molecular biologists from other institutions to scrutinize the genomes of SIVs from sooty mangabeys and rhesus macaques, and to create a provisional family tree. Their work, published in 1989 with Vanessa M. Hirsch as first author, revealed that SIV_{sm} is closely related to HIV-2. So is SIV_{mac}. "These results suggest that SIV_{sm} has infected macaques in captivity and humans in West Africa," the group wrote, placing the onus of origination on sooty mangabeys, "and evolved as SIV_{mac} and HIV-2, respectively." In fact, those three strains were very similar, suggesting divergence fairly recently from a common ancestor.

"A plausible interpretation of these data," Hirsch and her coauthors added, to make the point plainly, "is that in the past

30–40 years SIV from a West African sooty mangabey (or closely related species) successfully infected a human and evolved as HIV-2." It was official: HIV-2 is a zoonosis.

5

But what about HIV-1? Where did the great killer come from? That was a larger mystery that took somewhat longer to solve. The logical inference was that HIV-1 must be zoonotic in origin also. But what animal was its reservoir? When, where, and how had spillover occurred? Why had the consequences been so much more dire?

HIV-2 is both less transmissible and less virulent than HIV-1. The molecular bases for those fateful differences are still secrets embedded in the genomes, but the ecological and medical ramifications are clear and stark. HIV-2 is confined mostly to West African countries such as Senegal and Guinea-Bissau (the latter of which, during colonial times, was Portuguese Guinea), and to other areas connected socially and economically within the former Portuguese empire, including Portugal itself and southwestern India. People infected with HIV-2 tend to carry lower levels of virus in their blood, to infect fewer of their sexual contacts, and to suffer less severe or longer-delayed forms of immunodeficiency. Many of them don't seem to progress to AIDS at all. And mothers who carry HIV-2 are less likely to pass it to their infants. The virus is bad, but not nearly so bad as it could be. HIV-1 provides the comparison. HIV-1 is the thing that afflicts tens of millions of people throughout the world. HIV-1 is the pandemic scourge. To understand how the AIDS catastrophe has happened to humanity, scientists had to trace HIV-1 to its source.

This takes us to the city of Franceville, in southeastern Gabon, Central Africa, and an institution called the Centre International de Recherches Médicales (CIRMF), the same place at which important work on Ebola virus has been done. It's a nexus for the study of, and response to, emerging African diseases. At the end of the 1980s, a young Belgian woman named Martine Peeters worked as a research assistant at CIRMF for a year or so, during the period between getting her diploma in tropical medicine and going on for a doctorate. The CIRMF facility maintained a compound of captive primates, including three dozen chimpanzees, and Peeters along with several associates was tasked with testing the captive animals for antibodies to HIV-1 and HIV-2. Almost all of the chimps tested negative—all except two. Both exceptions were very young females, recently captured from the wild. Such baby chimps, like other orphan primates, are sometimes kept or sold off as pets after the killing and eating of their mothers. One of these animals, a two-year-old suffering from gunshot wounds, had been brought to CIRMF for medical treatment. She died of the wounds, but not before surrendering a blood sample. The other was an infant, maybe six months old, who survived. Blood serum from each of them reacted strongly when tested against HIV-1, less strongly when tested against HIV-2. That much was notable but slightly ambiguous. Antibody testing is an indirect gauge of infection, relatively convenient and quick, but imprecise. Greater precision comes with detecting fragments of viral RNA or, better still, isolating a virus—catching the thing in its wholeness and growing it in quantity—from which a confident identification can be made. Martine Peeters and her co-workers succeeded in isolating a virus from the baby chimp. Twenty years later, when I called on her at her office at an institute in southern

France, Peeters remembered vividly how that virus showed up in a series of molecular tests.

"It was especially surprising," she said, " because it was so close to HIV-1."

Had there been any previous hints?

"Yes. At that time we knew already that HIV-2 most likely came from primates in West Africa," she said, alluding to the sooty mangabey work. "But there was no virus close to HIV-1 already detected in primates. And until now, it's still the only virus close to HIV-1." Her group published a paper, in 1989, announcing the new virus and calling it SIV_{cpz}. They did not crow about having found the reservoir of HIV-1. Their conclusion from the data was more modest: "It has been suggested that human AIDS retroviruses originated from monkeys in Africa. However, this study and other previous studies on SIV do not support this suggestion." Left implicit: Chimpanzees, not monkeys, might be the source of the pandemic bug.

By the time I met her, Martine Peeters was director of research at the Institut de Recherche pour le Développement (IRD), in Montpellier, a handsome old city just off the Mediterranean coast. She was a small, blonde woman in a black sweater and silver necklace, concise and judicious in conversation. What sort of response met this discovery? I asked.

"HIV-2, people accepted it readily." They accepted, she meant, the notion of simian origins. "But HIV-1, people had more difficulties to accept it."

Why the resistance? "I don't know why," she said. "Maybe because we were young scientists."

The 1989 paper got little attention, which seems peculiar in retrospect, given the novelty and gravity of what it implied. In 1992 Peeters published another, describing a third case of SIV_{cpz}, this one in a captive chimpanzee that had been shipped to Brussels from Zaire. All three of her SIV-positive results were from "wild-born" chimpanzees taken captive (as distinct from animals bred in captivity) but that still left a gap in the chain of evidence. What about chimps still *in* the wild?

With only such tools of molecular biology as available in the early 1990s, the screening of wild chimps was difficult (and unacceptable to most chimp researchers), because the diagnostic tests required blood sampling. Lack of evidence from wild populations, in turn, contributed to skepticism in the AIDS-research community about the link between HIV-1 and chimps. After all, if Asian macaques had become infected with HIV-2 in their cages, from contact with African monkeys, might not SIV-positive chimpanzees simply reflect cage-contact infections too? Another reason for skepticism was the fact that, by the end of the 1990s, roughly a thousand captive chimpanzees had been tested but, apart from Peeters's three, not a single one had yielded traces of SIV_{cpz}. These two factors—the absence of evidence from wild populations and the extreme rarity of SIV in captive chimps—left open the possibility that both HIV-1 and SIV_{cpz} derived directly from a common ancestral virus in some other primate. In other words, maybe those three lonely chimps had gotten their infections from some still-unidentified monkey, and maybe the same unidentified monkey had given HIV-1 to humans. With that possibility dangling, the origin of HIV-1 remained uncertain for much of the decade.

In the meantime, researchers investigated not just the source of HIV but also its diversity in humans, discovering three major lineages of HIV-1. "Groups" became the preferred term for these lineages. Each group was a cluster of strains that was genetically discrete from the other clusters; there was variation *within* each group, since HIV is always evolving, but the differences *between* groups were far larger. This pattern of groups had some dark implications that dawned on scientists only gradually and still haven't been absorbed in the popular understanding of AIDS. I'll get to them shortly, but first let's consider the pattern itself.

Group M was the most widespread and nefarious. The letter M stood for "main," because that group accounted for most of the HIV infections worldwide. Without HIV-1 group M, there was no global pandemic, no millions of deaths. Group O was the second to be delineated, its initial standing for "outlier," because it encompassed only a small number of viral isolates, mostly traceable to what seemed an outlier area relative to the hotspots of the pandemic: Gabon, Equatorial Guinea, and Cameroon, all in western Central Africa. By the time a third major group was discovered, in 1998, it seemed logical to label that one N, supposedly indicating "non-M/non-O" but also filling in the alphabetical sequence. (Years later, a fourth group would be identified and labeled P.) Group N was extremely rare; it had been found in just two people from Cameroon. The rarity of N and O put group M dramatically in relief. M was everywhere. Why had that particular lineage of virus, and not the other two (or three), spread so broadly and lethally around the planet?

Parallel research on HIV-2, the less virulent virus, also found distinct groups but even more of them. Their labeling came from the beginning of the alphabet rather than the middle, and by the year 2000 seven groups of HIV-2 were known: A, B, C, D, E, F, and G. (An eighth group, turning up later, became H.) Again, most of them were extremely rare—represented, in fact, by viral samples taken from only one person. Groups A and B *weren't* rare; they accounted for the majority of HIV-2 cases. Group A was more common than group B, especially in Guinea-Bissau and Europe. Group B was traceable mainly to countries on the eastern end of West Africa, such as Ghana and Côte d'Ivoire. Groups C through H, although tiny in total numbers, were significant in showing a range of diversity.

As the new century began, AIDS researchers pondered this roster of different viral lineages: seven groups of HIV-2 and three groups of HIV-1. The seven groups of HIV-2, distinct as they were from one another, all resembled SIV_{sm}, the virus endemic in sooty mangabeys. (So did the later addition, group H.) The three kinds of HIV-1 all resembled SIV_{cpz}, from chimps. (The eventual fourth kind, group P, is most closely related to SIV from gorillas.) Now here's the part that, as it percolates into your brain, should cause a shudder: Scientists think that each of those twelve groups (eight of HIV-2, four of HIV-1) reflects an independent instance of cross-species transmission. Twelve spillovers.

In other words, HIV hasn't happened to humanity just once. It has happened at least a dozen times—a dozen that we know of, and probably many more times in earlier history. Therefore it wasn't a highly improbable event. It wasn't a singular piece of vastly unlikely bad luck, striking humankind with devastating results—like a comet come knuckleballing across the infinitude of space to smack planet Earth and extinguish the dinosaurs. No. The arrival of HIV in human bloodstreams was, on the contrary, part of a small trend. Due to the nature of our interactions with African primates, it seems to occur pretty often.

6

Which raises a few large questions. If the spillover of SIV into humans has happened at least twelve times, why has the AIDS pandemic happened only once? And why did it happen when it did? Why didn't it happen decades or centuries earlier? Those questions entangle themselves with three others, more concrete, less speculative, to which I already alluded: When, where, and how *did* the AIDS pandemic begin?

First let's consider when. We know from Michael Gottlieb's evidence that HIV had reached homosexual men in California by late 1980. We know from the case of Grethe Rask that it lurked in Zaire by 1977. We know that Gaëtan Dugas wasn't really Patient Zero. But if those people and places don't mark a real beginning point in time, what does? When did the fateful strain of virus, HIV-1 group M, enter the human population?

Two lines of evidence call attention to 1959.

In September of that year, a young print-shop worker in Manchester, England, died of what seemed to be immune-system failure. Because he had spent a couple years in the Royal Navy before returning to his hometown and his job, this unfortunate man has been labeled "the Manchester sailor." His health went into decline after his naval hitch, which he served mainly but not entirely in England. At least once, he sailed as far as Gibraltar. Back in Manchester by November 1957, he wasted away, suffering some of the symptoms later associated with AIDS, including weight loss, fevers, a nagging cough, and opportunistic infections, including *Pneumocystis carinii,* but no underlying cause of death could be determined by the doctor who did the autopsy. That doctor preserved some small bits of kidney, bone marrow, spleen, and other tissues from the sailor—embedding them in paraffin, a routine method for fixing pathology samples—and reported the case in a medical journal. Thirty-one years later, in the era of AIDS, a virologist at the University of Manchester tested some of those archived samples and found (or thought he found) evidence that the sailor had been infected with HIV-1. If that was correct, then the Manchester sailor would be recognized retrospectively as the first case of AIDS ever documented in the medical literature.

But wait. Retesting of the same samples by a pair of scientists in New York, several years later, showed that the earlier HIV-positive result must have reflected a laboratory mistake. The bone marrow now tested negative. The kidney material again tested positive but in a way that rang alarms of doubt: HIV-1 evolves quickly, and the genetic sequence of virus from the kidney sample seemed far too modern. It looked more like a modern variant than like something that could have existed in 1959. That suggested contamination with some recent strain of the virus to account for the positive tests. Conclusion: The Manchester sailor may have died from immune-system failure but HIV probably wasn't the cause. His case merely illustrates how tricky it can be to make a retrospective diagnosis of AIDS, even with the presence of what seems to be good evidence.

Soon after that false lead from Manchester was debunked, another lead emerged in New York. By now it was 1998. A team of researchers including Tuofu Zhu, based at the Rockefeller University, obtained an archival specimen from Africa dating back to the same year as the sailor's, 1959. This time it wasn't tissues; it was a small tube of blood plasma, drawn from a Bantu man in what had been Léopoldville, capital of the Belgian Congo (nowadays Kinshasa, capital of the Democratic Republic of the Congo), and stored for decades in a freezer. The man's name and his cause of death weren't reported. His sample had been screened during an earlier study, in 1986, along with 1,212 other plasmas—some archival, others new—from various locations in Africa. This man's was the only one that tested unambiguously positive for HIV. Tuofu Zhu and some colleagues probed further, working with what little remained of the original sample and using polymerase chain reaction (PCR, a biochemical technique that involves heating and cooling a sample in the presence of a certain enzyme) to amplify fragments of the viral genome.

Then they sequenced the fragments to assemble a genetic portrait of the man's virus. In their paper, published in February 1998, they called the sequence ZR59, referencing Zaire (as the country had long been known) and the year 1959. Comparative analysis showed that ZR59 was quite similar to both subtype B and subtype D (finer divisions within the HIV-1 group M lineage) but fell about halfway between, which suggested that it must closely resemble their common ancestor. In other words, ZR59 was a glimpse back in time, a genuinely old form of HIV-1, not a recent contamination. ZR59 proved that HIV-1 had been present—simmering, evolving, diversifying—in the population of Léopoldville by 1959. In fact it proved more. Further analysis of ZR59 and other sequences, led by Bette Korber of the Los Alamos National Laboratory, yielded a calculation that HIV-1 group M might have entered the human population around 1931.

For a decade, from the Zhu publication in 1998 until 2008, that landmark stood alone. ZR59 was the only known version of HIV-1 from a sample taken earlier than 1976. Then someone found another. This one became known as DRC60, and by now you can probably decode the label yourself: It came from the Democratic Republic of the Congo (same nation as Zaire, latest name, often abbreviated as DRC) and had been collected in 1960.

DRC60 was a biopsy specimen, a piece of lymph node snipped from a living woman. Like the Manchester sailor's bits of kidney and spleen, it had been locked away in a little pat of paraffin. Thus preserved, it needed no refrigeration, let alone freezing. It was as inert as a dead butterfly and less fragile. It could be stored and ignored on a dusty shelf—as it had been. After more than four decades, it emerged from a specimen cabinet at the University of Kinshasa and offered a new jolt of insight to AIDS researchers.

7

The University of Kinshasa sits on a hilltop near the edge of the city, reachable by an hour's taxi ride through the broken streets, the smoggy sprawl, the snarled traffic of vans and buses and pushcarts, past the street-side vendors of funerary wreaths, the cell-phone-recharge kiosks, the fruit markets, the meat markets, the open-air hardware stores, the tire-repair shops and cement brokers, the piles of sand and gravel and garbage, the awesome decrepitude of a postcolonial metropolis shaped by eight decades of Belgian opportunism, three decades of dictatorial misrule and egregious theft, then a decade of war, but filled with 10 million striving people, some of whom are dangerous thugs (as in all cities) and most of whom are amiable, hopeful, and friendly. The university campus, on its hill, loosely called "the mountain," presents a relatively verdant and halcyon contrast to the city below. Students go there, climbing by foot from a crowded bus stop, to learn and to escape.

Professor Jean-Marie M. Kabongo is head of pathology in the university's Department of Anatomic Pathology. He's a small, natty man with a huge graying handlebar mustache and full muttonchops, making a forceful visual impression that's vitiated by his gentle manner. He wears heavy brown glasses and a white lab coat over his shirt and tie. When I met the professor in his office, on the second floor of a building that overlooks a grassy concourse shaded with acacias, he pleaded imperfect knowledge of DRC60 and the patient from whom that specimen came. An old case, after all, going back long before his time. Yes, a woman, he believed. His memory was vague but he could check the records. He began taking notes as I questioned him and suggested I come back in a couple days, when he might be better prepared with answers. But then I asked about the room where DRC60 had been stored, and he brightened. Oh, of course, he said, I can show you *that*.

He fetched a key. Down a corridor beyond his office, he unlocked a blue door. Swinging it open, he welcomed me into a large sunlit laboratory with walls of white tile and two long, low tables down the middle. Against one wall stood an old GE refrigerator. On one of the low tables rested an old-fashioned folio ledger, with curling pages, like something from Chancery in the time of Dickens. On the far windowsill stood a row of beakers containing liquids in increments of color, beaker by beaker, from piss-yellow to vodka-clear. The yellowest, Professor Kabongo told me, was methanol. The clearest was xylol. We use these in preparing a tissue sample, he said. The point of such organic solvents is to extract the water; desiccation is prerequisite to fixing tissues for the long term. The methanol was darkened from processing many samples.

He showed me a small orange plastic basket, with a hinged lid, about the size and shape of a matchbook. This is a "cassette," Professor Kabongo explained. You take a lump of tissue from a lymph node or some other organ and enclose it in such a cassette; you soak the whole thing in the beaker of methanol; from the methanol, it goes through the intermediate baths in sequence; finally you dunk it in the xylol. Methanol draws out the water; xylol draws out the methanol, preparing your specimen for preservation in paraffin. And this device, Professor Kabongo said, indicating a large machine on one of the tables, delivers the paraffin. You take a leached tissue sample from its cassette, he explained, and place it on a little silver tray, here. From that spigot, you dribble out a stream of warm, liquid paraffin. It cools on the sample like a pat of butter. Now you remove the cassette lid and label the base with an individual code—for instance, A90 or B71. That's your archival specimen, he said. "A" means that it came from an autopsy. "B" indicates a biopsy. So the paraffin-caked bit of lymph node that yielded DRC60 would have been labeled B-something, as a biopsy specimen. Each coded specimen gets recorded in the big ledger. Then the specimens go into storage.

Storage. Storage where? I asked.

At the far end of the lab was another doorway, this one hung with a blue curtain. Professor Kabongo pushed the curtain aside and I followed him into a specimen pantry, narrow and tight, lined with shelves and cabinets along one side. The shelves and cabinets contained thousands of dusty paraffin blocks and old microscope slides. The paraffin blocks were in stacks and cartons, some of the cartons dated, some not. It appeared to be organized chaos. A wooden stool awaited use by any curious, tireless soul wishing to rummage through the samples. Although I didn't plan to rummage, my tour had suddenly come to its crescendo. *Here?* Yes, just here, said the professor. This is where DRC60 sat for decades. He could have added, with local pride: before becoming a Rosetta stone in the study of AIDS.

8

From the pantry behind the blue curtain, that sample and hundreds of others had traveled a circuitous route, to Belgium and then the United States, ending up in the laboratory of a young biologist at the University of Arizona. Michael Worobey is a Canadian, originally from British Columbia, whose specialty is molecular phylogenetics. After his undergraduate work he went to Oxford on a Rhodes scholarship, which ordinarily means two years of mildly strenuous academic work plus lots of tea, sherry, tennis on grass, and genteel anglophilia before the "scholar" returns to professional school or a career. Worobey put Oxford to more serious use, staying on, finishing a doctorate and then a postdoc fellowship in evolutionary biology at the molecular level. From there he returned to North America in 2003, accepting an assistant professorship at Arizona and building himself a BSL-3 lab (Biosafety Level 3, the second most secure category) for work on the genomes of dangerous viruses. Several years later, it was Worobey who detected evidence of HIV in a certain Congolese biopsy specimen from 1960.

Worobey amplified fragments of the viral genome, pieced the fragments together, recognized them as an early version of HIV-1, and named the sequence DRC60. Comparing his sequence with ZR59, the other earliest known strain, he reached a dramatic conclusion: that the AIDS virus has been present in humans for decades longer than anyone thought. The pandemic may have gotten its start with a spillover as early as 1908.

To appreciate Worobey's discovery and how it splashed down amid previous ideas, you need a little context. That context involves a heated dispute over just how HIV-1 entered the human population. The prevailing notion as of the early 1990s, based on what had been learned about HIV-2 and the sooty mangabey, among other factors, was that HIV-1 also came from an African primate, and that it had probably gotten into humans by way of two separate instances (for groups M and O, the ones then recognized) of butchering bushmeat. This became known as the cut-hunter hypothesis. In each instance, a man or a woman had presumably butchered the carcass of an SIV-positive primate and suffered exposure through an open wound—maybe a cut on the hand, or a scratch on the arm, or a raw spot on any skin surface that got smeared with the animal's blood. A wound on the back might have sufficed, if the carcass were draped over shoulders for carrying home. A wound in the mouth, if some of the meat were consumed raw. All that mattered was blood-to-blood contact. The cut-hunter hypothesis was speculative but plausible. It was parsimonious, requiring few complications and no unlikelihoods. It fit the known facts, though the known facts were fragmentary. And then in 1992 a contrary theory arose.

This one was heterodox and highly controversial: that HIV-1 first got into humans by way of a contaminated polio vaccine tested on a million unsuspecting Africans. The vaccine itself, by this theory, had been an unintended delivery system for AIDS. Someone, according to the theory, had monumentally goofed. Someone was culpable. Scientific hubris had overridden caution, with horrendous results. The scariest thing about the polio-vaccine theory was that it also seemed plausible.

Viruses are subtle. They get in where they shouldn't. Laboratory contaminations occur. Even viral or bacterial contamination of a vaccine at the production level—it has happened. Back in 1861, a group of Italian children vaccinated against smallpox, with material direct from a "vaccinal sore," came down with syphilis. Smallpox vaccine administered to kids in Camden, New Jersey, at the start of the twentieth century, seems to have been contaminated with tetanus bacillus, resulting in the death of nine vaccinated children from tetanus. Around the same time, a batch of diphtheria antitoxin prepared in St. Louis, using blood serum from a horse, also turned out to carry tetanus, which killed another seven children. Producers then began filtering vaccines, an effective precaution against bacterial contamination; but viruses passed through the filters. Formaldehyde was sometimes added to inactivate a target virus, and that supposedly killed unwanted viruses too, but the supposition wasn't always correct. As late as the midcentury, some of the early batches of the Salk polio vaccine were adulterated with a virus known as SV40, endemic in rhesus macaques. SV40 in vaccine became a hot issue, several years later, when suspicions arose that the virus causes cancer. It's important to note that vaccine purification methods and safety precautions have improved vastly since the time of the Salk vaccine, and that adamant resistance to vaccination by some activists in the present era is misguided, unjustified by scientific data, and costly to public health. But a half century ago, the possibility of vaccine contamination did exist.

Whether vaccine contamination happened with HIV-1, around the same time as the SV40 problem and far more consequentially, is another matter. That the vaccine in question had been given to Africans was not in dispute. Between 1957 and 1960, a Polish-born American researcher named Hilary Koprowski—a lesser-known competitor in the same vaccine-development race that engaged Jonas Salk and Albert Sabin—arranged for his oral polio vaccine to be widely administered in areas of the eastern Belgian Congo and adjacent colonial holdings. These were parts of what would eventually be DRC, Rwanda, and Burundi. Koprowski himself visited Stanleyville, in 1957, and made contacts who later oversaw the trials. Children and adults lined up trustingly, in places like the Ruzizi Valley north of Lake Tanganyika, to receive doses of liquid vaccine from a tablespoon or a squirting pipette. Spritz, you're good. Next! The numbers are uncertain. By one account, roughly seventy-five thousand kids were vaccinated just in Léopoldville. The heterodox theory argued two additional points about this enterprise: First, that Koprowski's vaccine was produced by growing the virus in chimpanzee kidney cells (rather than in monkey kidney cells, the standard technique); second, that at least some batches of that vaccine were produced from chimpanzee kidneys drawn from animals infected with SIV_{cpz}.

The result of that flawed vaccinating, certain people have argued, was iatrogenic infection (disease caused by medical treatment) of an unknown number of Central Africans with what later became recognized as HIV-1. By this notion, known for short as the OPV (oral polio vaccine) theory, a single reckless researcher had seeded the continent—and the world—with AIDS.

The OPV theory has been around and notorious since 1992, when a freelance journalist named Tom Curtis described it in a long article for *Rolling Stone.* Curtis's piece ran under the headline: THE ORIGIN OF AIDS: A STARTLING NEW THEORY ATTEMPTS TO ANSWER THE QUESTION, "WAS IT AN ACT OF GOD OR AN ACT OF MAN?" Several other researchers had mooted the idea earlier, more obscurely, and one of them had put Tom Curtis onto the story. When Curtis started looking into it, some eminent scientists responded with defensive dismissals, which served only to suggest that maybe the theory did merit consideration. Curtis even drew a brusque and ill-considered comment from the head of research for the World Health Organization's Global Programme on AIDS, Dr. David Heymann: "The origin of the AIDS virus is of no importance to science today." He quoted another expert, William Haseltine of Harvard, as saying: "It's distracting, it's nonproductive, it's confusing to the public, and I think it's grossly misleading in terms of getting to the solution of the problem." After publication of the piece, lawyers for Hilary Koprowski filed a lawsuit against Curtis and *Rolling Stone,* charging defamation, and the magazine ran a "clarification," admitting that the OPV theory and Koprowski's role represented just an unsupported hypothesis. But as the dust settled at *Rolling Stone,* an English journalist named Edward Hooper took hold of the OPV theory as a personal obsession and an investigative crusade, giving it a second life.

Hooper spent years researching the subject with formidable tenaciousness (though not always critical good sense) and in 1999 made his case in a thousand-page book titled *The River: A Journey to the Source of HIV and AIDS.* Hooper's river was a metaphorical one: the flow of history, the stream of cause-and-effect, from a very small beginning to an ocean of consequences. In the book's prologue, he alluded to the quest by Victorian explorers for the source of the Nile. Does that river begin from Lake Victoria, pouring out at Ripon Falls, or is there another and more obscure source upstream from the lake? "The controversy surrounding the source of the Nile," Hooper wrote, "is strangely echoed by another controversy of a century and a half later, the long-running debate about the origins of AIDS." The Victorian explorers had been wrong about the Nile and, according to Hooper, so were the modern experts wrong about the starting point of the AIDS pandemic.

Hooper's book was massive, overwhelmingly detailed, seemingly reasonable, exhausting to plod through but mesmerizing in its claims, and successful at bringing the OPV theory to broader public attention. Some AIDS researchers (including Phyllis Kanki and Max Essex) had long been aware that vaccine contamination, with SIV from monkey cells, was at least a theoretical possibility; they had even conducted screening efforts on vaccine lines, and found no evidence of such a problem. Hooper, following Tom Curtis, raised the idea from a concern to an accusation. His vast river of information and his steamboat of argument didn't prove the essential thesis—that Koprowski's vaccine had been made from chimp cells contaminated with HIV. But his work did seem to raise the possibility that the vaccine *could* have been made from chimp cells that *might* have been contaminated.

The issue of possibility then gave way to the issue of fact. What had actually happened? Where was the evidence? At the urging of an eminent evolutionary biologist named William Hamilton, who believed that the OPV theory deserved investigation, the Royal Society convened a special meeting in September 2000 to discuss the subject within its broader context. Hamilton was a senior figure, liked and respected, whose early work in evolutionary theory helped inform Edward O. Wilson's *Sociobiology* and Richard Dawkins's *The Selfish Gene*. He swung the Royal Society into giving the OPV theory a fair hearing. Edward Hooper, though not a scientist himself, was invited to speak. Hilary Koprowski also came, as well as a roster of leading AIDS researchers. By the time that meeting convened, though, William Hamilton was dead.

He died suddenly in March 2000, of intestinal bleeding, after an attack of malaria contracted during a research trip to the Democratic Republic of the Congo. In his absence, his colleagues at the Royal Society discussed a wide range of matters related to the origins of HIV and AIDS. The OPV theory was just one topic among many, though implicitly it drove the agenda of the whole meeting. Did the available data from molecular biology and epidemiology tend to support, or to refute, the vaccine-contamination scenario? A corollary to that question was: When had HIV-1 first entered the human population? If the earliest infections occurred before 1957, those infections couldn't have resulted from Koprowski's OPV trials. Archival HIV-positives might therefore be decisive.

This is the context that brought DRC60 out of Kinshasa. A senior Congolese virologist named J. J. Muyembe, aware of the archived pathology specimens at the University of Kinshasa and equally aware of the OPV debate, took it upon himself to enlarge the body of available data. Muyembe went up to the university, rifled through the pantry behind the blue curtain, packed 813 paraffin-embedded specimens into an ordinary suitcase, and carried it with him on his next professional visit to Belgium. There he handed the trove to a colleague named Dirk Teuwen, who had taken part in the Royal Society meeting a couple years earlier. Teuwen, in accord with a prior agreement for collaborative study, sent them to Michael Worobey in Tucson.

These two lines of narrative fold back into each other. Worobey, as a grad student, knew both Bill Hamilton at Oxford and some of the disease biologists in Belgium. Impelled by his own interest in the origins of HIV, Worobey accompanied Hamilton to Congo on that last fatal fieldtrip. They went in January 2000, during the chaotic aftermath of a civil war, which had replaced the longtime potentate Mobutu Sese Seko with the upstart Laurent Kabila as president of the DRC. Hamilton wanted to collect fecal and urine samples from wild chimpanzees; those specimens, he hoped, might help confirm or refute the OPV theory. Worobey, for his part, put little stock in the OPV theory but wanted more data from which to chart the origin and evolution of HIV. It was a crazy time in the Democratic Republic of the Congo, more crazy than usual, because two rebel armies opposed to Laurent Kabila still controlled much of the eastern half of the country. Hamilton and Worobey flew into Kisangani (formerly Stanleyville), a regional capital along the upper Congo River, the same city where Koprowski had begun his Congo enterprise. Now it was occupied by Rwanda-backed forces on one riverbank and Uganda-backed forces on the other. Commercial airlines weren't flying, because of the war, so the two biologists shared a small, chartered plane with a diamond dealer. In Kisangani they paid their respects to the Rwanda-backed commander, whose ambit included most of the city, and as quickly as possible got out into the forest, where they would be safer among the leopards and snakes. They spent a month collecting fecal and urine samples from wild chimpanzees, with help from local guides, and by the time they left, Hamilton was sick.

Neither he nor Worobey knew *how* sick, but they caught the next exit flight they could, which took them to Rwanda. From there they bounced to Entebbe in Uganda, where Hamilton got a confirmed diagnosis of falciparum malaria and some treatment, then onward to Nairobi, and from Nairobi up to London Heathrow. By now Hamilton seemed past the worst of his illness; he was feeling much better. They had accomplished their mission and life was good. An American field biologist once expressed to me how he felt in such moments. "That's the name of the game: getting home with the data." This man's research too involved dangers—shipwreck, starvation, drowning, snakebite, though not malaria and Kalashnikov rifles. "If you take too many risks, you don't get home," he said. "If you take too few, you don't get the data." Hamilton and Worobey got the data, got home, then learned that the ice cooler containing their precious chimp specimens had gone astray in luggage handling somewhere between Nairobi and London.

I visited Michael Worobey in Tucson to hear about all this. "Everything was fine," he told me, "except we checked six bags, including the cooler that had samples, and five of our bags came through the carousel and the one with the samples disappeared." His friend Hamilton, feeling ill again the next morning, went to a hospital—and hemorrhaged disastrously, perhaps due to anti-inflammatory drugs he'd been taking against the malarial fever. Worobey phoned and got the news from Hamilton's sister: *Who are you why are you calling Bill is in extremis*. Worobey meanwhile had been hassling by long-distance phone with a luggage handler in Nairobi, who assured him that the cooler had been found and would arrive on the next flight. What arrived was someone else's cooler—full of sandwiches or somesuch, as he recalled. "So that was an extra bit of drama that unfolded as Bill was dying in the hospital," Worobey told me. The correct cooler arrived two days later but Hamilton was in no shape to celebrate. He went through a series of surgeries and transfusions and then, after weeks of struggle, he died.

The fecal samples from Congolese chimps, for which Hamilton had given his life, yielded no SIV-positives. A couple of urine samples registered in the borderline zone for antibodies. Those results weren't clear or dramatic enough to merit publication. Good data are where you find them, not always where you look. Several years later, when the human pathology samples from Kinshasa reached Tucson—those 813 little blocks of tissue in paraffin, the ones J. J. Muyembe had carried to Belgium in a suitcase—Michael Worobey was ready. He found DRC60 among them, and it told an unexpected story.

9

Screening paraffin-embedded hunks of old organ samples to find viral RNA isn't easy, not even for an expert. Those little blocks, Worobey said, turned out to be "some of the nastiest kinds of tissues to do molecular biology with." The problem wasn't forty-three years at room temperature in a dusty equatorial pantry. The problem was the chemicals used in fixing the tissues—the 1960 equivalent of the beakers of methanol and xylol that Professor Kabongo had shown me. Back in those days, pathologists favored something called Bouin's fixative, a potent little mixture containing mostly formalin and picric acid. It worked well for preserving the cellular structure of tissues, like salmon in aspic, so that samples could be sliced thin and examined under a microscope; but it was hell on the long molecules of life. It tended to break up DNA and RNA into tiny fragments, Worobey explained, and form new chemical bonds, leaving "sort of a big, tangled mess rather than a nice string of beads that you can do molecular biology on." Because the process was so laborious, he screened just 27 of the 813 tissue blocks from Kinshasa. Among those twenty-seven, he found one containing RNA fragments that unmistakably signaled HIV-1. Worobey persisted adeptly, untangling the mess and fitting the fragments to assemble the sequence of nucleotide bases he named DRC60.

That was the wet work. The dry work, done largely by computer, entailed base-by-base comparisons between DRC60 and ZR59. It also involved broader comparisons, placing those two within a family tree of known sequences of HIV-1 group M. The point of such comparisons was to see how much evolutionary divergence had occurred. How far apart had these strains of virus grown? Evolutionary divergence accumulates by mutation at the base-by-base level (other ways too, but those aren't relevant here), and among RNA viruses such as HIV, the mutation rate is relatively fast. Equally important, the average rate of HIV-1 mutation is known—or anyway, it can be carefully estimated from the study of many strains. That rate of mutation is considered the "molecular clock" for the virus. Every virus has its own rate, and therefore its own clock measuring the ticktock of change. The amount of difference between two viral strains can therefore reveal how much time has passed since they diverged from a common ancestor. Degree of difference factored against clock equals elapsed time. This is how molecular biologists calculate an important parameter they call TMRCA: time to most recent common ancestor.

Okay so far? You're doing great. Take a breath. Now those bits of understanding will boost us across a deep gulf of molecular arcana to an important scientific insight. Here we go.

Michael Worobey found that DRC60 and ZR59, sampled from people in Kinshasa during almost the same year, were *very* different. They both fell within the range of what was unmistakably HIV-1 group M; neither could be confused with group N or group O, nor with the chimp virus, SIV_{cpz}. But within M, they had diverged *far*. How far? Well, one section of genome differed by 12 percent between the two versions. And how different was that, measured in time? About fifty years' worth, Worobey figured. More precisely, he placed the most recent common ancestor of DRC60 and ZR59 in the year 1908, give or take a margin of error.

Spillover back in 1908? That's much earlier than anyone suspected, and therefore the sort of discovery that gets into an august journal such as *Nature*. Publishing in 2008, a century after the fact, with a list of coauthors that included J. J. Muyembe and Jean-Marie Kabongo, Worobey wrote:

> Our estimation of divergence times, with an evolutionary timescale spanning several decades, together with the extensive genetic distance between DRC60 and ZR59 indicate that these viruses evolved from a common ancestor circulating in the African population near the beginning of the twentieth century.

To me he said: "This wasn't a new virus in humans."

Worobey's work directly refuted the OPV hypothesis. If

HIV-1 existed in humans as early as 1908, then obviously it hadn't been introduced via vaccine trials beginning in 1958. Clarity on that point was valuable—but only a side benefit of Worobey's contribution. Even more important was the basic fact of his early date. Placing the crucial spillover in time, so long ago, represented a big step toward understanding how the AIDS pandemic may have started and grown.

10

Placing the spillover in *space* was equally important, and achieved by a different laboratory. Beatrice Hahn is somewhat older than Worobey and had begun her work on the origin of AIDS long before he found DRC60.

Born in Germany, Hahn took a medical degree in Munich, then came to the United States in 1982 and spent three years as a postdoc in Robert Gallo's lab, studying retroviruses. She moved next to the University of Alabama at Birmingham, where she became Professor of Medicine and Microbiology and codirector of a center for AIDS research, with a group of bright postdocs and grad students working under her aegis. (She remained at Alabama from 1985 to 2011, a period encompassing most of the work described here, and then joined the Perelman School of Medicine at the University of Pennsylvania, in Philadelphia.) The broader purpose of Hahn's various projects, and the goal she shares with Worobey, is to understand the evolutionary history of HIV and its relatives and antecedents. The fittest label for that sort of research is the one Worobey mentioned when I asked him to describe his field: molecular phylogenetics. A molecular phylogeneticist scrutinizes the nucleotide sequences in the DNA or RNA of different organisms, comparing and contrasting, for the same reason a paleontologist scrutinizes fragments of petrified bone from extinct giant saurians—to learn the shape of lineages and the story of evolutionary descent. But for Beatrice Hahn especially, as a medical doctor, there's an additional purpose: to detect how the genes of HIV function in causing disease, toward the prospects of better treatment, prevention, and maybe even a cure.

Some very interesting papers have come out of Hahn's laboratory in the past two decades, many of them published with a junior researcher as first author and Hahn in the lab leader's position, last. That was the case in 1999, when Feng Gao produced a phylogenetic study of SIV_{cpz} and its relationship to HIV-1. At the time there were only three known strains of SIV_{cpz}, all drawn from captive chimps, with Gao's paper adding a fourth. The work appeared in *Nature,* highlighted by a commentary calling it "the most persuasive evidence yet that HIV-1 came to humans from the chimpanzee, *Pan troglodytes.*" In fact, Gao and his colleagues did more than trace HIV-1 to the chimp; their analysis of viral strains linked it to individuals of a particular subspecies known as the central chimpanzee, *Pan troglodytes troglodytes*, whose SIV had spilled over to become HIV-1 group M. That subspecies lives only in western Central Africa, north of the Congo River and west of the Oubangui. So the Gao study effectively identified both the reservoir host and also the geographical area from which AIDS must have arisen. It was a huge discovery, as reflected in the headline of *Nature*'s commentary: FROM *PAN* TO PANDEMIC. Feng Gao at the time was a postdoc in Hahn's lab.

But because Gao based his genetic comparisons (as Martine Peeters had done earlier) on viruses drawn from captive chimps, the soupçon of uncertainty about infection among wild chimpanzees remained, at least for a few more years. Then, in 2002, Mario L. Santiago topped a list of coauthors announcing in *Science* their discovery of SIV_{cpz} in the wild. Santiago was a PhD student of Beatrice Hahn's.

The most significant aspect of Santiago's work, for which he got his richly deserved doctorate, was that on the way toward detecting SIV in a single wild chimpanzee (just one animal among fifty-eight tested), he invented methods by which such detections could be made. The methods were "noninvasive," meaning that a researcher didn't need to capture a chimp and draw its blood. The researcher needed only to follow animals through the forest, get under them when they pissed (or, better still, send a field assistant into that yellow shower), collect samples in little tubes, and then screen the samples for antibodies. Turns out that urine could be almost as telling as blood.

"That was a breakthrough," Hahn told me, during a talk at her lab in Birmingham. "We weren't sure it would work." But Santiago took the risk, cooked up the techniques, and it did work. The very first sample of SIV-positive urine from a wild chimpanzee came from the world's most famous community of chimps: the ones at Gombe National Park, in Tanzania, where Jane Goodall had done her historic field study, beginning back in 1960. That trace of virus didn't match quite as closely with HIV-1 as Feng Gao's had done, and it came from a chimp of a different subspecies, the eastern chimpanzee, *Pan troglodytes schweinfurthii*. But it was SIV_{cpz} nonetheless.

The advantage of sampling at Gombe, Hahn told me, was that those chimps didn't run away. They were truly wild but, after four decades of study by Goodall and her successors, well habituated to human presence. For use elsewhere, the urine-screening method wasn't practical. "Because, you know, non-habituated chimps don't stay close enough so you can catch their pee." You could collect their poop from the forest floor, of course, but fecal samples were useless unless preserved somehow; fresh feces contain an abundance of proteases, digestive enzymes, which would destroy the evidence of viral presence long before you got to your laboratory. These are the constraints within which a molecular biologist studying wild animals labors: the relative availability and other parameters of blood, shit, and piss.

Another of Hahn's young wizards, Brandon F. Keele, soon solved the problem of fecal sample decay. He did it by tinkering with a liquid stabilizer called RNAlater, a commercial product made by a company in Austin, Texas, for preserving nucleic acids in tissue samples. The nice thing about RNAlater is that its name is so literally descriptive: The stuff allows you to retrieve RNA from a sample . . . later. If it worked with RNA in tissues, Keele reasoned, maybe it could work also with antibodies in feces. And indeed it did, after he and his colleagues untangled the chemical complications of getting those antibodies released from the fixative. This technique vastly enlarged the scope of screening that was possible on wild chimpanzees. Field assistants could collect hundreds of fecal samples, scooping each into a little tube of RNAlater, and those samples—stored without refrigeration, transported to a distant laboratory—would yield their secrets later. "If we find the antibodies, we know that chimps are infected," Hahn told me. "And then we can home in on those we know are infected, and try to get the viruses out." Antibody screening is easy and quick. Performing PCR amplification and the other requisite steps to probe for fragments of viral RNA is far more laborious. The new methods allowed Hahn and her group to look first at a large number of specimens and then work more concertedly on a select few. They could separate the Shinola from the shit.

And they could expand their field surveying beyond Gombe.

They could turn their attention back to *Pan troglodytes troglodytes,* the subspecies of chimp whose SIV_{cpz} most closely matched HIV-1. Working now with Martine Peeters of Montpellier, plus some contacts in Africa, they collected 446 samples of chimpanzee dung from various forest sites in the south and southeast of Cameroon, after which Brandon Keele led the laboratory analysis. DNA testing showed that almost all the samples came from *P. t. troglodytes* (though a couple dozen derived from a different chimp subspecies, *P. t. vellerosus*, whose range lay just north of a major river). Keele then looked for evidence of virus. The samples yielded two surprising results.

11

To hear about those surprises, I visited Brandon Keele, who by this time had finished his postdoc with Hahn and gone off to a research position at a branch of the National Cancer Institute, in Frederick, Maryland. He was still studying viral phylogenetics and AIDS, as head of a unit devoted to viral evolution. His new office and lab were on the grounds of Fort Detrick, a high-security installation that once housed the U.S. biological weapons program and still encompasses USAMRIID, the big army research institute on infectious diseases. Since I was entering without an escort, soldiers at the guardhouse searched the underside of my rental car for a bomb before letting me pass. Keele, waiting to flag me down outside the door of his building, wore a blue dress shirt, jeans, his black hair moussed back, and a two-day stubble. He is a tall young man, extremely polite, raised and educated in Utah. We sat in his small office and looked at a map of Cameroon.

The first surprise to emerge from the fecal samples was high prevalence of SIV_{cpz} in some communities of Cameroonian chimps. Two that scored highest, Keele said, were at sites labeled Mambele (near a crossroads by that name) and Lobeke (within a national park). Whereas all other sampling of chimps had suggested that SIV infection was rare, the sampling in southeastern Cameroon showed prevalence rates up to 35 percent. But even there, the prevalence was "spotty," Keele said. "We can sample hundreds of chimps at a site and find nothing." But go just a little farther east, cross a certain river, sample again, and the prevalence spikes upward. That was unexpected. The rates were especially high in the farthest southeastern corner of the country, where two rivers converge, forming a wedge-shaped national boundary. This wedge of Cameroon appears to jab down into the Republic of the Congo (not to be confused with the DRC), its neighbor to the southeast. The wedge was a hotspot for SIV_{cpz}.

The second surprise came once he extracted viral fragments from the samples, amplified those fragments, sequenced them, and fed the genetic sequences into a program that would compare these new strains with many other known strains of SIV and HIV. The program expressed its comparisons in the form of a most-probable phylogeny—a family tree. Keele recalled watching the results for a certain chimp, an individual labeled LB7, whose feces had been collected at Lobeke. "We were just shocked," he said. "I mean, I had ten people around my computer, all waiting to see what that sequence looked like." What it looked like was the AIDS virus.

When his computer delivered its latest tree, LB7's isolate of SIV_{cpz} showed up as a twig amid the same little branch that held all known human strains of HIV-1 group M. (In scientific lingo, it fell within the same *clade.*) It was at that point "the closest thing" to a match, Keele told me, that had ever been found in a wild chimp. "And then we find more, right? The more we dig, the more we find." The other close matches came from that same little area: southeastern Cameroon. A chilling, historic epiphany, at which Keele and his colleagues were thrilled. "You can't make this stuff up, as Beatrice would say. It's too good." Their joy lasted about ten seconds, after which everyone became hungry for more samples and more results. Your celebration is always provisional, Keele told me, until you've written the paper and gotten that congratulatory note of acceptance from the editors of *Science.*

Keele and the group now sequenced entire genomes (not just fragments) from four samples, all collected in the same area, and on those sequences ran their genetic analyses again. Again they found the new SIV_{cpz} shockingly similar to HIV-1 group M. The similarity was so close as to leave almost no chance that any other variant, yet undiscovered, could be much closer. Hahn's lab had located the geographical origin of the pandemic: southeastern Cameroon.

12

So much for *where* as well as *when*. AIDS began with a spillover from one chimp to one human, in or near that small southeastern wedge of Cameroon, around 1908 (give or take a margin of error). From there it grew, slowly but inexorably, from a spillover to an outbreak to a pandemic. That leaves our third question: *how?*

The Keele paper appeared in *Science*, on July 28, 2006, under the title "Chimpanzee Reservoirs of Pandemic and Nonpandemic HIV-1." In addition to Brandon Keele as first author, there was the usual list of coauthors, including Mario Santiago, Martine Peeters, several partners from Cameroon, and last again, Beatrice H. Hahn. The data were fascinating, the conclusions were judicious, the language was careful and tight. Near the end, though, the authors let supposition fly:

> We show here that the $SIV_{cpz\mathit{Ptt}}$ strain that gave rise to HIV-1 group M belonged to a viral lineage that persists today in *P. t. troglodytes* apes in southeastern Cameroon. That virus was probably transmitted locally. From there it appears to have made its way via the Sangha River (or other tributaries) south to the Congo River and on to Kinshasa where the group M pandemic was probably spawned.

But the phrase "transmitted locally" was opaque. What mechanism, what circumstances? How did those crucial events occur and proceed?

Hahn herself, along with three coauthors, had addressed that back in 2000, when she first argued the idea that AIDS is a zoonosis: "In humans, direct exposure to animal blood and secretions as a result of hunting, butchering, or other activities (such as consumption of uncooked contaminated meat) provides a plausible explanation for the transmission." She was alluding to the cut-hunter hypothesis. More recently she addressed it again: "The likeliest route of chimpanzee-to-human transmission would have been through exposure to infected blood and body fluids during the butchery of bushmeat." A man kills a chimpanzee and dresses it out, hacks it up, in the course of which he suffers blood-to-blood contact through a cut on his hand. SIV_{cpz} passes across the species boundary, from chimp to human, and taking hold in the new host becomes HIV. This event is unknowable in its particulars but it's plausible, and it fits the established facts. Some variant of the cut-hunter scenario, occurring in a forest of southeastern Cameroon around 1908, would account not just for Keele's data but also for Michael Worobey's timeline. But then what? One man in southeastern Cameroon is infected.

"If the spillover occurred there," I asked Hahn, "how was it that the epidemic began in Kinshasa?"

"Well, there are lots of rivers going down from that region to Kinshasa," she said. "And the speculation, the hypothesis, is that is how the virus traveled—in people, not in apes. It wasn't the apes that got into the canoe for a little visit of Kinshasa. It was the people who carried the virus down, most likely." Sure, she acknowledged, there was a slim chance that someone might have brought a live chimp, captive, infected, all the way down from the Cameroonian wedge—"but I think it is highly unlikely." More likely the virus traveled in humans.

Sexual contacts in the villages kept the chain of infection alive, though barely, by this line of speculation, and the disease didn't explode as a notable outbreak—not for a long while. When someone died of immunodeficiency, the death may have seemed unremarkable amid all other sources of mortality. Life was hard, life was perilous, life expectancy was short even apart from the new disease, and many of those earliest HIV-positive people may have succumbed to other causes before their immune systems failed. There was no epidemic. But the chain of infection sustained itself. Each HIV-positive person infected, on average, at least one other person. The virus seems to have traveled just as people traveled in those days: mainly by river. It made its way out of southeastern Cameroon along the headwaters of the Sangha River, then down the Sangha to the Congo, then down the

Congo to Brazzaville and Léopoldville, the two colonial towns on either side of a huge broadening of the river, which was then known as the Stanley Pool. "Once it got into an urban population," Hahn said, "it had an opportunity to spread."

But still it moved slowly, like a locomotive just leaving the station. Léopoldville contained fewer than ten thousand people in 1908, and Brazzaville was even smaller. Sexual mores and the fluidity of interactions were unlike what prevailed in the boondocks, but not yet so unlike as they would become. The arithmetic of the outbreak was still modest, with scarcely more than a single transmission onward per infected person. In the language of disease ecologists: the basic reproduction rate (the average number of secondary infections from each primary infection at the beginning) barely exceeded 1.0, the minimal level required for the outbreak to continue indefinitely. Then, as time passed, more people drifted into the towns, drawn by the prospect of working for wages or selling their goods. Habits and opportunities changed. Women came as well as men, though not so many of them, and among those who did, more than a few entered the sex trade.

By 1914, Brazzaville contained about six thousand people and was "a hard mission field," according to one Swedish missionary, where "hundreds of women from upper Congo are professional prostitutes." It was the capital of the colony then known as French Equatorial Africa. The male population included French civil servants within the colonial administration, soldiers, traders, and laborers, and they probably outnumbered females by a sizable margin, due to colonial policies that discouraged married men, coming there to work, from bringing their families. That gender imbalance heightened the demand for commercial sex. But the format for bought favors, in those early years, was generally different from what the word "prostitute" might suggest—grindingly efficient, wham-bam encounters with a long succession of strangers. Instead there were single women, known as *ndumbas* in Lingala and *femmes libres* in French, "free women" as distinct from wives or daughters, who would provide their clients with a suite of services, ranging from conversation to sex to washing clothes and cooking. One such ndumba might have just two or three male friends who returned on a regular basis and kept her solvent. Another variant was the *ménagère*, a "housekeeper" who lived with a white colonial official and did more than keep house. Commercial arrangements, yes, but these didn't represent the sort of prodigiously interconnected promiscuity that could cause a sexually transmitted virus to spread widely.

Across the pool in Léopoldville, meanwhile, the disparity of genders was even worse. This town was essentially a labor camp, controlled by its Belgian administrators, inhospitable to families, where the male-female ratio in 1910 was ten to one. Travel through the countryside and entry into Léopoldville were restricted, especially for adult females, though some women managed to get false documents or evade the police. If you were a restless, imaginative girl in one of the villages, poorly fed and poorly treated, to be a ndumba in Léopoldville could well have seemed enticing. Here too, though, even with ten horny men for each woman, commercial sex didn't happen in brothels or by streetwalking. Free women had their special friends, their clients, maybe several contemporaneously, but there was no dizzying permutation of multiple sexual contacts, not yet. One expert has called this "a low-risk type of prostitution," with regard to the prospects of HIV transmission.

Léopoldville also supported a lively market in smoked fish. Ivory, rubber, and slaves were traded there, for export, with profits going mainly to white concessionaires, well into the colonial era. Although a deep canyon and a set of forbidding cataracts stood between the Stanley Pool and the river's mouth, isolating both cities from the Atlantic Ocean, a portage railway built in 1898 had breached that isolation, bringing more goods and commerce, which brought more people, and in 1920 Léopoldville replaced a downriver town as capital of the Belgian Congo. By 1940, its population had edged up to forty-nine thousand. Then the demographic curve steepened. Between 1940 and independence, which came in 1960, the city grew by almost an order of magnitude, to about four hundred thousand people. Léopoldville became Kinshasa, a twentieth-century African metropolis, where life was very different from what passed in a Cameroonian village. The tenfold population increase, along with the concomitant changes in social relations, might go a long way to explain why HIV "suddenly" took off. By 1959, the ZR59 carrier was infected, and a year later in the same city the carrier of DRC60 too. By that time the virus had proliferated to such a degree, mutating and diversifying, that DRC60 and ZR59 represented quite different strains. The basic reproduction rate now must have well exceeded 1.0, and the new disease spread—through the two cities and eventually beyond. "You know," Hahn said, "a virus was at the right place at the right time."

When I read Keele's presentation of the chimp data and the analysis, in early 2007, my jaw dropped like a pound of ham. These folks had located Ground Zero, if not Patient Zero. And when I looked at the map—Figure 1 in Keele's paper, showing the Cameroonian wedge and its surroundings—I saw places I knew. A village where I had slept, on my way to a Congo assignment for *National Geographic*. A river I had ascended in a motor pirogue. It turned out that, seven years earlier, during an arduous footslog expedition through the forests of the Republic of the Congo and Gabon, with an American ecologist named J. Michael Fay, he and I and his forest crew had passed very near the cradle of AIDS.

After talking with Beatrice Hahn, I thought it might be illuminating to go back.

13

Douala is a city on Cameroon's western coast, with a seaport and an international airport. I escaped it, with my own crew, as quickly as possible. We rode east in a beat-up but sturdy Toyota truck, leaving at dawn, getting ahead of the crush, our gear stashed under tarps in the pickup's bed. Moïse Tchuialeu was my driver, Neville Mbah my Cameroonian fixer, and Max Mviri, from the Republic of the Congo, was along to handle things when we reentered his country in the course of the crazy loop I had planned. Max and I had flown up from Brazzaville the night before. We were a genial foursome, eager to move after the hassles of preparation, rolling past the closed shops and the billboards to the city's eastern fringe, where traffic thickened in a haze of blue diesel exhaust and the outlier markets were already open for business, selling everything from pineapples to phone minutes. Highway N3 would take us straight to Yaoundé, Cameroon's inland capital, and then another big two-lane onward from there.

During a stop in Yaoundé, around midday, I met with a man named Ofir Drori, head of an unusual group called LAGA (the Last Great Ape Organization) that helps government agencies in Central Africa enforce their wildlife-protection laws. I wanted to see Drori because I knew that LAGA was especially engaged on the problem of apes' being killed for bushmeat. I found him to be a lean Israeli expat with dark, alert eyes and a patchy goatee. Wearing a black shirt, black jeans, a black ponytail, and an earring, he looked like a rock musician or, at least, a hip New York waiter. But he seemed to be a serious fellow. He had come to Africa as an adventure-seeking eighteen-year-old, Drori told me, and gotten involved with human-rights work in Nigeria, then moved to Cameroon, did a little gorilla journalism (or was it *guerrilla* journalism?), and became a passionate antipoaching organizer. He founded LAGA, he said, because enforcement of Cameroon's antipoaching laws had been terrible, nonexistent, for years. The group now provided technical support to investigations, raids, and arrests. Subsistence hunting for duikers and other abundant, unprotected kinds of animal is legal in Cameroon, but apes, elephants, lions, and a few other species were protected by law—and increasingly by actual enforcement. Perpetrators were finally getting busted, even doing time, for dealing in ape flesh and other contraband wildlife products. Drori gave me a LAGA newsletter describing the efforts to stem poaching of chimps and gorillas, and he warned me against the myth that ape hunting is a problem because local people are hungry. The reality, he said, is that local people eat duikers or rats or squirrels or monkeys—if they eat meat at all—whereas the fancy stuff, the illicit delicacies, the chimpanzee body parts, the gobs of elephant flesh, the hippopotamus steaks, get siphoned away by upscale demand from the cities, where premium prices justify the risks of poaching and illegal transport. "What brings the money are the protected species," he said. "Things that are rare."

Drori's newsletter mentioned a raid against a hidden storage room, at a train station, that served at least three different dealers; the room had contained six refrigerators and its seized contraband included a chimpanzee hand. Another bust, against a dealer whose car had held fifty kilos of marijuana plus a young chimp with a bullet wound, suggested diversified wholesale commerce. And if chimp meat travels toward money, chimp viruses presumably do too. "If you're thinking about infection," he said, knowing that I was, "don't just think of villages." Any chimpanzee killed in the southeastern corner of the country, including an SIV-positive individual, might easily end up here in Yaoundé, being sold for meat in a back alley or served through a very discreet restaurant.

We left the city in early afternoon, headed eastward again, moving against a stream of log trucks hammering toward us in the opposite lane, each one burdened to capacity with a load of just five or six gigantic stems. Somewhere out there, in that sparsely populated corner of the country, old-growth forests were being sheared. Around sundown we reached a town called Abong Mbang and stopped at the best local hotel, which meant running water and a lightbulb. Early next day, an hour beyond Abong Mbang, the blacktop ended though the log trucks kept coming, now on a ribbon of red clay. The temperature climbed toward midday equatorial heat and, wherever we encountered a little rain shower, the road steamed in red. Elsewhere the landscape was so dry that powdery red clay dust rose on the gusts from passing vehicles, coating trees along the roadside like bloody frost. We hit a police checkpoint and endured a routine but annoying shakedown, which Neville handled with aplomb, making two phone calls to influential contacts, refusing to pay the expected bribe, and yet somehow recovering our passports after only an hour. This guy is good, I thought. The road narrowed further, to a band of arsenical red barely wider than a log truck, leaving us hugging the shoulder when we encountered one, and the forest thickened on both sides. Around noon we crossed the Kadéï River, greenish brown and slow, meandering southeast, a reminder that we were now at the headwaters of the Congo basin. The villages through which we passed became smaller and looked progressively more spare and poor, with few gardens, little livestock, almost nothing for sale except bananas, mangoes, or a bowl of white manioc chips set out forlornly on an untended stand. Occasionally a goat or a chicken scampered out of our way. In addition to the log trucks, we now met flatbeds loaded with milled lumber, and I remembered hearing how such trucks sometimes carried a concealed stash of bushmeat, rumbling toward the black markets of Yaoundé and Douala. (A photographer and activist named Karl Ammann documented that tactic with a photo, taken at a road junction here in southeastern Cameroon, of a driver unloading chimpanzee arms and legs from the engine compartment of his log truck. The photo appeared in a book by Dale Peterson, titled *Eating Apes*, in which Peterson estimated that the human population of the Congo basin consumes roughly 5 million metric tons of bushmeat each year. Much of that wild meat—though no one knows just how much—travels out of the forest as contraband cargo on log trucks.) Apart from the trucks, today on this stretch of red clay, there was almost no traffic. By late afternoon we reached Yokadouma, a town of several thousand. The name translates as "Fallen Elephant," presumably marking the site of a memorable kill.

We found a local office of the World Wildlife Fund and, inside, two earnest Cameroonian employees named Zacharie Dongmo and Hanson Njiforti. Dongmo showed me a digital map plotting the distribution of chimpanzee nests in this southeastern corner of the country, which includes three national parks—Boumba Bek, Nki, and Lobeke. A chimpanzee nest is simply a small platform of interwoven branches, often in the fork of a smallish tree, which provides just enough support for the ape to sleep comfortably. Each individual makes one each night, though a mother will share hers with an infant. Tallying such nests, which remain intact for weeks after a one-night use, is how biologists estimate chimpanzee populations. The pattern on Dongmo's map was clear: a high density of nests (and therefore of chimpanzees) within the parks, a low density outside the parks, and none at all in areas adjacent to the roads leading to Yokadouma. Logging and bushmeat were the reasons. Logging operations bring roads and workers and firearms into the depths of the forest; dead wildlife consequently travels out. Dongmo and Njiforti explained it as an informal, ad hoc form of commerce. "Most of the illegal trade is man-to-man," Njiforti said. "A poacher meets you and says, I have meat." But it's also woman-to-man, he added: Much of the trading is done by "Buy 'em–Sell 'ems," women who travel between villages as petty traders, dealing openly in cloth, or spices, or other staples, and covertly in bushmeat. Such a woman buys directly from the hunter, often paying in bullets or shotgun shells, and sells to whomever she can. Commerce is relatively fluid; many of these women have cell phones. And there are all sorts of tricks, Njiforti said, for getting the meat out. It could be tucked into a truckload of cocoa pods, for instance, a cash crop from this region. The police and the wildlife wardens may get tipped off, and they can stop a truck and search it, but at some risk to themselves. If you stop a truck and demand it be unloaded, and then there's no illegal cargo, Njiforti said, "the guy can sue you. The information has to be very good." That's why Ofir Drori's network has proved itself useful.

Most of the poachers, Dongmo added, are Kakos, a tribe from the north with a strong affinity for bushmeat. Many of them have drifted down here to the southeast, drawn by marital connections or opportunity in the bush. The local Baka Pygmies, on the other hand, have traditional strictures against eating apes, which are deemed too close to human. In fact, Dongmo reckoned, there was probably less eating of apes down here than in some other sectors of the country—apart from the totemic consumption of ape parts by Bakwele people in connection with a certain initiation ceremony for adolescent boys. And that offhanded comment from Zacharie Dongmo was the first I'd heard of a Bakwele ritual known as *beka*.

We lingered in Yokadouma for two nights and a day, long enough for me to walk the dirt streets, admire the concrete statue of an elephant gracing the town's central roundabout, photograph a piteous pangolin about to be butchered for meat, and encounter a fellow who told me more about beka. This man, whose name I omit, had written a small report on the subject, which his organization declined to publish. He gave me a copy. Yes, he said, the Bakwele people here in the southeast use chimp and gorilla meat in their beka ceremony. They especially favor the arms. As a result, he said, "chimps are becoming more and more scarce." So scarce that gorilla arms are now often used as a substitute.

His report described a typical beka initiation, complete with slaughtered sheep and chickens, the neck of a tortoise (because it resembles a penis), and "virgin lasses" in attendance through a long prelude that culminates at four in the morning. The boy to be initiated is dressed in leaves and given drugs to keep him awake. Drums beat through the night until, before dawn, the boy is led into a special area of forest, where he's obliged to confront two chimpanzees. Some of what follows seems to be symbolic enactment, some of it blood-real. "A gong is sounded," according to a Bakwele chief who informed my source, "a voice calls out from the forest, and two chimpanzees respond. The male chimpanzee comes out first and touches the boy's head. The female chimpanzee emerges minutes after and the boy is expected to kill it." At dawn the boy bathes, then stays awake until late afternoon, pacing and expectant, at which point the circumciser comes at him with a homemade knife. "I nursed my wound for 45 days after," one initiate said. But now he was a man, no longer a boy. The unpublished report added:

> Until recently, the Bakweles have been using chimps for this ritual. They claim two chimps could be used for circumcision of as many as 36 people. They amputate the arms of the chimps. This part of the animal is eaten by elders of the village. Of late, however, due to the scarcity of chimps, Bakweles go for gorillas.

Eight gorilla arms had recently been seized from a poacher who fled when confronted by game rangers, leaving the meat behind in a bag. The arms were intended for an impending beka. "We cannot do without these animals," the Bakwele chief complained, "if we must perform this important traditional rite."

It's no condescension against Bakwele culture to note that butchering chimpanzees to eat their arms as part of an ancient and bloody ritual could be a very good way to acquire SIV_{cpz}. Then again, in a landscape as lean and severe as southeastern Cameroon in 1908, beka might have been superfluous. Sheer hunger could account for the original spillover just as well.

14

Thirty miles farther south, at a crossroads known as Mambele Junction, with a central roundabout defined by three truck tires piled up like coins, we dined by kerosene lantern at a small cantina, eating smoked fish (at least, I hoped it was smoked fish) in peanut sauce and drinking warm Muntzig beer. This happened to be the place where Karl Ammann saw chimpanzee arms stashed under the hood of a log truck. It was also one of the locations featured in Brandon Keele's paper on the chimpanzee origins of HIV-1. Chimp fecal samples from hereabouts had shown high prevalence of the virus in its most fateful form. Somewhere very nearby was Ground Zero of the AIDS pandemic.

After dinner, my compadres and I stepped back outside and admired the sky. Although this was Saturday night, the lights of Mambele Junction didn't amount to much and despite their dim glow we could see not just the Big Dipper, Orion's Belt, and the Southern Cross but even the Milky Way, arcing overhead like a great smear of glitter. You know you're in the boonies when the galaxy itself is visible downtown.

Two days later, at a modest building nearby that served as headquarters for Lobeke National Park, I met with the park's *conservateur*, its director, a handsomely bald man named Albert Munga, dressed in a floral shirt and (unmatched) floral pants. He sat aloof at his desk for several minutes, shuffling papers, before deigning to notice me, and then for a while longer he seemed cool to my questions about chimpanzees. The office was heavily air-conditioned; everything about it was cool. But after half an hour Mr. Munga warmed, loosened, and began to share some of his data and his concerns. The park's population of great apes (chimps and gorillas combined) had fallen abruptly since 2002, he told me: from about sixty-three hundred animals to about twenty-seven hundred. Commercial poachers were the problem, and by his account they came mainly across the eastern boundary of the park, the Sangha River, which happens also to be the southeastern border of Cameroon. Beyond the Sangha lie the Central African Republic and, slightly farther south, the Republic of the Congo, two countries that have known insurgency and war in the last couple decades. Those political conflicts have brought military weapons (especially Kalashnikov rifles) into the region, vastly increasing the difficulty of protecting animals. Bands of well-armed poachers come across the river, mow down elephants and anything else they see, whack out the ivory and the elephant meat, lop off the heads and limbs of the apes, take the smaller creatures whole, and escape back across the water. Or else they move their booty downriver by boat. "There is a huge bushmeat traffic on the Sangha," Munga told me, "and the destination is Ouesso." The town of Ouesso, a river port of some twenty-eight thousand people, just over the border in Congo, is a major trading nexus on the upper Sangha. By no coincidence, it was my destination too.

Just outside Mr. Munga's office, I paused in the corridor to look at a wall poster with lurid illustrations and a warning in French: *LA DIARRHEA ROUGE TUE!* Red diarrhea kills. At first glance I thought that referred to Ebola virus disease, another gruesome affliction faced by Central African villagers; but no. "*Grands Singes et VIH/SIDA,*" read the finer print. VIH is the French equivalent of HIV, and SIDA likewise is AIDS. The cartoonish but unfunny drawings depicted a stark parable about the connection between simian bushmeat and death. I lingered long enough for the oddness to strike me. Throughout the rest of the world you see AIDS-education materials crying out: *Practice safe sex! Wear a condom! Don't share needles!* Here the message was: *Don't eat apes!*

We drove onward, along a dirt track between walls of green, still farther into Cameroon's southeastern wedge. The country's southern border out here is formed by the Ngoko River, a tributary flowing east to its junction with the Sangha. The Ngoko, according to local lore, is one of the deepest rivers in Africa, but if so there must be a steep wrinkle of rock underneath, because it's only about eighty yards wide. We reached it around midday at a town called Moloundou, a scruffy place spread over small hills above the river. From any good point of vantage in Moloundou, the Republic of the Congo was easily visible across the water—close enough that, in the quiet of evening, we could hear the chainsaws of illegal loggers at work in the darkness over there. These log poachers would fell trees directly into the water and tangle them into rafts, I was told, then float the rafts down to Ouesso, where a mill operator would pay cash, no questions asked. Ouesso again: the outlaw entrepôt. There was no government presence, no law, no timber concessionaires defending their interests, on that side—so said scuttlebutt on this side, anyway. We had reached the frontier zone, which was still a bit wild and woolly.

Early next morning we walked up to the market and watched sellers setting out their goods in neat piles and rows: local peanuts and pumpkinseeds and red palm nuts, garlic and onions, manioc tubers, plantains, giant snails and smoke-blackened fish, hocks of meat. I hung back discreetly from the meat counters, leaving Neville and Max to investigate what was available. Mostly it was smoked duiker, a form of small forest antelope that served as a staple wild food; no sign of ape meat being sold aboveboard; and even pangolin, a seller told Neville, was out of season. I hadn't expected different. Anything so valuable as a chimpanzee carcass would change hands in private, probably by prior arrangement, and not be slabbed out at a public market.

Downstream from Moloundou, the last Cameroonian outpost on the Ngoko River is Kika, a logging town with a big mill that provides jobs and lodging for hundreds of men and their families, plus a dirt airstrip for the convenience of its managerial elite. There was no direct riverside road (why would there be? the river *is* a road) so we circled back inland to get there. Arriving in Kika, we reported promptly to the police station, a small shack near the river that served also as immigration post, where an officer named Ekeme Justin roused himself, pulled on his yellow T-shirt, and performed the necessary formalities for Max and me: stamping our passports *sortie de Cameroon.* We would exit the country here. Officer Justin, upon receipt of a fee for his stamp work, became our great friend and host, offering us tent space there beside the police post and help in finding a boat. He went off to town with Neville, the all-purpose fixer, and by sunset they had arranged charter of a thirty-foot wooden pirogue, with an outboard, capable of getting Max and me across the river border and down to Ouesso.

I was up at 5:10 the next morning, packing my tent, eager to turn the corner on this big loop and head back into Congo. Then we waited through a heavy morning rain. Finally came our boatman, a languid young man named Sylvain in a green tracksuit and flip-flops, to mount his outboard and bail the pirogue. We loaded, covered our gear with a tarp against the lingering drizzle, and after warm good-byes to the faithful Neville and Moïse, also Officer Justin, we launched, catching a strong current on the Ngoko. We pointed ourselves downriver. For me, of course, this journey was all about the cut-hunter hypothesis. I wanted to see the route HIV-1 had traveled from its source and imagine the nature of its passage.

15

Let's give him due stature: not just a cut hunter but the Cut Hunter. Assuming he lived hereabouts in the first decade of the twentieth century, he probably captured his chimpanzee with a snare made from a forest vine, or in some other form of trap, and then killed the animal with a spear. He may have been a Baka Pygmy man, living independently with his extended family in the forest or functioning as a sort of serf under the "protection" of a Bantu village chief. But probably he wasn't, given what I had heard of Baka scruples about eating ape. More likely he was Bantu, possibly of the Mpiemu or the Kako or one of the other ethnic groups inhabiting the upper Sangha River basin. Or he may have been a Bakwele, involved in the practice of beka. There's no way of establishing his identity, nor even his ethnicity, but this remote southeastern corner of what was then Germany's Kamerun colony offered plenty of candidates. I imagine the man being thrilled and a bit terrified when he found a chimpanzee caught in his snare. He had proved himself a successful hunter, a provider, a proficient member of his little community—and he wasn't yet cut.

The chimp too, tethered by a foot or a hand, would have been terrified as the man approached, but also angry and strong and dangerous. Maybe the man killed it without getting hurt; if so, he was lucky. Maybe there was an ugly fight; he might even have been pummeled by the chimp, or badly bitten. But he won. Then he would have butchered his prey, probably on the spot (discarding the entrails but not the organs, such as heart and liver, which were much valued) and probably with a machete or an iron knife. At some point during the process, perhaps as he struggled to hack through the chimp's sternum or disarticulate an arm from its socket, the man injured himself.

I imagine him opening a long, sudden slice across the back of his left hand, into the muscular web between thumb and forefinger, his flesh smiling out pink and raw almost before he saw the damage or felt it, because his blade was so sharp. And then immediately his wound bled. By a lag of some seconds, it also hurt. The Cut Hunter kept working. He'd been cut before and it was an annoyance that barely dimmed his excitement over the prize. His blood flowed out and mingled with the chimp's, the chimp's flowed in and mingled with his, so that he couldn't quite tell which was which. He was up to his elbows in gore. He wiped his hand. Blood leaked again into his cut, dribbled again into it from the chimp, and again he wiped. He had no way of knowing—no language of words or thoughts by which to conceive—that this animal was SIV-positive. The idea didn't exist in 1908.

The chimpanzee's virus entered his bloodstream. He got a sizable dose. The virus, finding his blood to be not such a different environment from the blood of a chimp, took hold. *Okay, I can live here.* It did what a retrovirus does: penetrated cells, converted its RNA genome into double-stranded DNA, then penetrated further, into the cells' nuclei, and inserted itself as DNA in the DNA genome of those host cells. Its primary targets were T cells of the immune system. A certain protein receptor (CD4) on the surface of those cells, in the Cut Hunter, was not very different from the equivalent receptor (another CD4) on the T cells of the butchered chimpanzee. The virus attached, entered the human cells, and made itself at home. Once integrated into the cellular genome, it was there for good. It was part of the program. It could proliferate in two ways: by cell replication (each time an infected T cell copied itself, the retroviral genome was copied also) and by activating its little subgenome to print off new virus particles, which then escaped from the T cell and floated off to attack other cells. The Cut Hunter was now infected, though apart from a slash on the hand he felt fine.

Forget about Gaëtan Dugas. This man was Patient Zero.

Maybe he carried the chimp carcass, or parts of it, back to his village in triumph. Maybe, if he was Baka, he delivered the whole thing to his Bantu overlord. He didn't want to eat it anyway. If he was Bantu himself, his family and friends feasted. Or maybe the chimp was a windfall from which he could afford to take special profit. If the season had been bounteous, with some duikers or monkeys killed, some forest fruits and tubers to eat, a good crop of manioc, so that his family wasn't starving, he may have lugged his chimpanzee to a market, like the one in Moloundou, and traded for cash or some valuable item, such as a better machete. In that case, the meat would have been parceled out retail and many people may have eaten bits of it, either roasted or smoked or dried. But because of how the virus generally achieves transmission (blood-to-blood or sexually) and how it doesn't (through the gastrointestinal tract), quite possibly none of those people received an infectious dose of virus, unless by contact of raw meat with an open cut on the hand or a sore in the mouth. A person might swallow plenty of HIV particles but, if those virions are greeted by stomach acids and not blood, they would likely fail to establish themselves and replicate. (The greatest danger, left unspecified in that "Don't eat apes" poster I saw, is not in the swallowing but in the killing, handling, and butchering of the animal.) Let's suppose that fifteen different consumers partook of the chimp meat, and that they all remained fine. HIV-negative. Lucky folks. Let's suppose that only the Cut Hunter became infected directly from the chimp.

Time passed. The virus abided and replicated within him. His infectiousness rose high during the first six months, as virions in multitude bloomed in his blood; then the viremia declined some as his body mounted an early immune response, while it still could, and leveled off, for a period of time. He noticed no effects. He passed the virus to his wife, eventually also to one of the four other women with whom he had sex. He suffered no immunodeficiency—not yet. He was a robust, active fellow who continued to hunt in the forest. He fathered a child. He drank palm wine and laughed with his friends. And then a year later, let's say, he died violently in the course of an elephant hunt, an activity even more perilous than butchering chimpanzees. He was one of seven men, all armed with spears, and the wounded elephant chose him. He took a tusk through the stomach, momentarily pinning him to the ground. You could see the tusk hole in the dirt afterward, as though a bloody stake had been driven in and pulled. Of the men who scooped him up, the women who prepared him for burial, none had an open cut and so they were spared infection. His son was born HIV-negative.

The Cut Hunter's widow found a new man. That man was circumcised, free of genital sores, and lucky; he didn't become infected. The other woman who had been infected by the Cut Hunter took several partners. She infected one. This fellow was a local chief, with two wives and occasional access to young village daughters; he infected both wives and one of the girls. The chief's wives remained faithful to him (by constraint of circumstance if not by choice), infecting no one. The infected girl eventually had her own husband. And so, onward. You get the idea. Although sexual transmission of the virus occurred less efficiently from female to male, and not all so efficiently from male to female, it was just efficient enough. After a few years, a handful of people had acquired the virus. And then still more, in time, but not many. Social life was constrained by small population size, absence of opportunity, and to some degree mores. The virus survived with a basic reproduction rate barely above 1.0. It passed to a second village, in the course of neighborly interactions, and then a third, but it didn't proliferate quickly in any of them. No one detected a wave of inexplicable deaths. It smoldered as an endemic infection at low prevalence in the populace of that little wedge of terrain, between the Ngoko River and the upper Sangha, where life tended to be short and hard.

People died young from all manner of mishaps and afflictions. If a young man, HIV-positive, was killed in a fight, no one knew anything about his blood status except that it had been spilled. If a young woman, HIV-positive, died of smallpox during a local outbreak, likewise she left no unusual story.

In some cases, during those early years, an infected person may have lived long enough to suffer immune failure. Then there were plenty of ready bugs, in the forest, in the village, to kill him or her. That wouldn't have seemed remarkable either. People died of malaria. People died of tuberculosis. People died of pneumonia. People died of nameless fever. It was routine. Some of those people might have recovered, had their immune systems been capable, but no one noticed a new disease. Or if someone did notice, the report hasn't survived. This thing remained invisible.

Meanwhile the virus itself may have adapted, at least slightly, to its new host. It mutated often. Natural selection was at work. Given a marginal increase in its capacity to replicate within human cells, leading to increased levels of viremia, its efficiency of transmission may have increased too. By now it was what we would call HIV-1 group M. A human-infecting pathogen, rare, peculiar, confined to southeastern Cameroon. Maybe a decade went by. The bug survived. Spillovers of SIV_{cpz} into humans had almost certainly occurred in the past (plenty of chimps were butchered, plenty of hunters were cut) and resulted in previous chains of infection, but those chains had been localized and short. The smoldering outbreak had always come to a cold end. This time it didn't. Before such burnout could occur, another person entered the situation—also hypothetical but fitted to the facts—whom I'll call the Voyager.

The Voyager wasn't a hunter. Not an expert and dedicated one, anyway. He had other skills. By my imagining, he was a fisherman. He lived not in a forest clearing like the one at

Mambele but in a fishing village along the Ngoko River. I picture him as a river boy from childhood. He knew the water; he knew boats. He owned a canoe, a good one, sturdy and long, made from a mahogany log with his own hands, and he spent his days in it. He was a young man with no wife, no children, and just a bit of an appetite for adventure. He had fallen away from his natal community at an early age, becoming a loner, because his father died and the village came to despise his mother, suspecting her of sorcery based on a piece of bad luck and a grudge. He took this as a deep personal bruise; he despised the villagers in return, screw them, and went his own way. It suited him to be alone. He was not an observant Bakwele. He never got circumcised.

The Voyager ate fish. He ate little else, in fact, besides fish and bananas—and sometimes manioc, which he didn't plant or process himself but which was easily bought with fish. He liked the taste and he loved the idea of fish, and there was always enough. He knew where to find fish, how to catch them, their varied types and names. He drank the river. That was enough. He didn't make palm wine or buy it. He was self-sufficient and contained within his small world.

He provided fish to his mother and her two younger children, as I see him, a loyal son though an alienated neighbor. His mother still lived at the fringe of the old village. His surplus catch he dried on racks, or in wet season smoked over a fire, at his solitary riverbank camp. Occasionally he made considerable journeys, paddling miles upstream or drifting downstream, to sell a boatload of fish in one of the market villages. In this way, he had tasted the empowerment of dealing for cash. Brass rods were the prevailing currency, or cowrie shells, and sometimes he may even have seen deutschmarks. He bought some steel hooks and one spool of manufactured line, which had come all the way from Marseille. The line was disappointing. The hooks were excellent. Once he had floated downstream as far as the confluence with the Sangha, a much larger river, powerful, twice as wide as the Ngoko, and had ridden its current for a day—a heady and fearful experience. On the right bank he had seen a town, which he knew to be Ouesso, vast and notorious; he gave it a wide berth, holding himself at midriver until he was past. At day's end he stopped, slept on the bank, and next day reversed, having tested himself enough. It took him four days of anxious effort to paddle back up, hugging the bank (except again at Ouesso), climbing through eddies, but the Voyager made it, relieved when he regained his own world, the little Ngoko River, and swollen with new confidence by the time he beached at his camp. This might have occurred, let's say, in the long dry season of 1916.

On another occasion, he paddled upstream as far as Ngbala, a river town some miles above Moloundou. It was during his return from that journey, as I posit, that he stopped at Moloundou and there, in his boat, where it was tied for the night in a shaded cove just below town, had sex with a woman.

She wasn't his first but she was different from village girls. She was a river trader herself, a Buy 'em–Sell 'em, several years older than he was and considerably more experienced. She traveled up and down the Ngoko and the Sangha, making a living with her wits and her wares and sometimes her body. The Voyager didn't know her name. Never heard it. She was outgoing and flirtatious, almost pretty. He didn't think much about pretty. She wore a print dress of bright calico, manufactured, not local raffia. She must have liked him, or at least liked his performance, because she returned to his boat in the shadows the next night and they coupled again, three times. She seemed healthy; she laughed merrily and she was strong. He considered himself lucky that night—lucky to have met her, to have impressed her, to have gotten at no cost what other men paid for. But he wasn't lucky. He had a small open wound on his penis, barely more than a scratch, where he'd been caught by a thorny vine while stepping ashore from a river bath. No one can know, not even in this imagined scenario, whether the lack of circumcision was crucial to his susceptibility, or the little thorn wound, or neither. He gave the woman some smoked fish. She gave him the virus.

It was no act of malice or irresponsibility on her part. Despite swollen and aching armpits, she had no idea she was carrying it herself.

16

River travel through tropical jungle is uncommonly soothing and hypnotic. You watch the walls of greenery slide by and, unless the channel is narrow enough for tsetse flies to notice your passing and come out from the shores, you suffer almost none of the discomforts. Because the riverbanks represent forest edges, admitting the full blast of sunlight, as closed canopy does not, the vegetation is especially tangled and rife: trees draped with vines, understory impenetrable, thick as the velvet curtain at an old movie theater. It presents an illusion that the forest itself, its interior, might be as dense as a sponge. But to a river traveler that density is immaterial because you have your own open route down the middle. If you've walked the forest, which is difficult though not sponge-thick, river journeying is an escape from impediments that feels almost akin to flight.

For a while after leaving Kika, we favored the Congo side, riding a strong channel. Sylvain knew his preferred line. His assistant, a Baka man named Jolo, handled the outboard while Sylvain supervised, signaling directions from the bow. The pirogue was large and steady enough that Max and I could sit on the gunnels. Immediately we passed a small police post on the right bank, a Congolese counterpart to the Cameroonian one at Kika, and fortunately no one flagged us to stop. Every such checkpoint in Congo is an occasion for passport-stamping and minor shakedowns, and you want to avoid them when you can. Then we puttered past a few villages, widely spaced, each just a cluster of wattle-and-daub houses sited on a high bank to escape inundation in wet season. The houses were topped with thatch and surrounded by banana trees, an oil palm or two, children in rag dresses and shorts. The kids stood transfixed as we passed. How many hours to our destination? I asked Sylvain. Depends, he said. Ordinarily he would stop in villages along the way for trading or passengers, delaying long enough to enter Ouesso by darkness so as to escape notice by the immigration police. Not long after that explanation he did stop, guiding us ashore at a village on the Congo bank, to which he delivered a large plastic tarp and from which, on departure, we gained a passenger.

It was my charter but I didn't mind. She was a young woman carrying two bags, an umbrella, a purse, and a pot of lunch. She wore an orange-and-green dress and a bandanna. I might have guessed if I hadn't been told: She was a Buy 'em–Sell 'em. Her name was Vivian. She lived down in Ouesso and would be glad for the ride home. She was lively and plump, confident enough to be traveling the river alone, trading in rice, pasta, cooking oil, and other staples. Sylvain liked to give her a lift because she was his sister—a statement of status that could be taken literally or not. She might have been his girlfriend or his cousin. Beyond this, I didn't learn much from Vivian except that her niche still exists, the Buy 'em–Sell 'em role, offering independent-spirited women a form of autonomy not easily found within village life, or even town life, and that the river still functions as a conduit of economic and social fluidity. Mostly she seemed a charming throwback and, though this might be unfair to her, put me in mind of women that the Voyager might have met almost a century earlier. She was a potential intermediary.

When the rain returned, Max and I and Sylvain and Vivian hunkered beneath our tarp, heads down but peeking out, while Jolo the Baka stolidly motored us on. We passed a solitary fisherman in his canoe, pulling a net. We passed another village from which children stared. Then the rain died again and the storm breeze fell off; the gentle chop disappeared, leaving the river as flat and brown as a cooled café au lait. Mangroves reached out from the banks like groping octopuses. I noticed a few egrets but no kingfishers. In midafternoon we approached the confluence with the Sangha. Along the left bank, the land fell gradually lower and then tapered, sinking away into the water. The Sangha River gripped us, swung us around, and I turned to watch that southeastern wedge of Cameroon recede to a vanishing point. The cradle of AIDS.

The air warmed slightly with an upstream breeze. We passed a large, wooded island. We passed a man standing upright in his dugout, paddling carefully. And then in the distance ahead, through haze, I saw white buildings. White buildings meant bricks and whitewash and governmental presence in something larger than a village: Ouesso.

Within half an hour we landed at the Ouesso waterfront, with its concrete ramp and wall, where an officer from the immigration police and a gaggle of tip-hungry, scuttering porters awaited. Stepping ashore, we had reentered the Republic of the

Congo. We completed the immigration formalities in French and then Max dealt with the bag-grabbing porters in Lingala. Sylvain and Jolo and Vivian melted away. Max was a shier, less forceful fellow than Neville but conscientious and earnest, and now it was his turn to be my fixer. He made some inquiries here along the waterfront and soon had good news: that the big boat, the cargo-and-passenger barge known as *le bateau*, would be departing tomorrow for Brazzaville, many miles and days further downriver. I wanted us to be on it.

We found a hotel, Max and I, and in the morning walked to the Ouesso market, which was centered in a squat, pagoda-shaped building of red brick just blocks from the river. The pagoda was big and stylish and old, with a concrete floor and a circular hall beneath three tiers of corrugated metal roof, dating back at least to colonial times. The market had far outgrown it, sprawling into a warren of wood-frame stalls and counters with narrow lanes between, covering much of a city block. Business was brisk.

A study of bushmeat traffic in and around Ouesso, done in the mid-1990s by two expat researchers and a Congolese assistant, had found about 12,600 pounds of wild harvest passing through this market each week. That total included only mammals, not fish or crocodiles. Duikers accounted for much of it and primates were second, though most of the primate meat was monkey, not ape. Eighteen gorillas and four chimps had been butchered and sold during the four-month study. The carcasses arrived by truck and by dugout canoe. As the biggest town in northern Congo, with no beef cattle to be seen, Ouesso was draining large critters out of the forest for many miles around.

Max and I snooped up and down the market aisles, stepping around mud holes, dodging low metal roofs, browsing as we had done in Moloundou. Because this was Ouesso, the merchandise was far more abundant and diverse: bolts of colorful cloth, athletic bags, linens, kerosene lanterns, African Barbie dolls, hair falls, DVDs, flashlights, umbrellas, thermoses, peanut butter in bulk, powdered fufu (instant manioc paste, just add water and stir for half an hour with a stick) in piles, mushrooms in buckets, dried shrimp, wild fruits from the forest, freshly fried beignets, blocks of bouillon, salt by the scoop, blocks of soap, medicines, bins of beans, pineapples and safety pins and potatoes. On one counter a woman hacked at live catfish with a machete. Just across from her, another woman offered a selection of dead monkeys. The monkey seller was a large middle-aged lady, her hair in cornrows, wearing a brown butcher's apron over her paisley dress. Genial and direct, she slapped a smoked monkey down proudly in front of me and named her price. Its face was tiny and contorted, its eyes closed, its lips dried back to reveal a deathly smile of teeth. Split up the belly and splayed flat, it was roughly the size and shape of a hubcap. *Six mille francs*, she said. Beside the first monkey she tossed down another, in case I was particular. *Six mille* for that one too. She was talking in CFA, the weak Central African currency. Her six thousand francs amounted to thirteen dollars, and was negotiable, but I passed. She also had a smoked porcupine, five duikers, and another simian, this one so freshly killed that its fur was still glossy and I could recognize its species: *Cercopithecus nictitans*, the greater spot-nosed monkey. That's a premium item, Max said, it'll go fast. Nearby, gobbets of smoked pork from a red river hog were priced at three thousand francs per kilo. All these animals could be hunted legally (though not with snares) and traded openly in Congo. There was no sign of apes. If you want chimpanzee or gorilla meat in Ouesso it can still be had, no doubt, but you've got to make private arrangements.

Our trip downriver on the bateau suffered complications and

delays so that, four days later, Max and I were back in Ouesso. Revisiting the market, we passed again through the pagoda, down the narrow aisles between stalls, along the counters piled with catfish and monkeys and duikers, smoked and fresh. This time I noticed a wheelbarrow full of smallish crocodiles and saw one croc being whacked apart on a plank. You could locate the meat section from anywhere in the market maze, I realized, by that sound—the steady *thunk-thunk!* of machetes. And then we came again to the brown-aproned lady, who remembered me. "You've returned," she said in French. "Why don't you buy something?" This time she plunked down a little duiker, more as a challenge than as an offering: *Are you a shopper or a voyeur?* I prefer chicken, I said lamely. Or smoked fish. Unsurprised by the pusillanimity of the white man, she smiled and shrugged. Then, as a flyer, I said: But if you had *chimpanzee* . . . She ignored me.

Or *elephant*, Max added. Now she laughed noncommittally and turned back to her real customers.

17

The idea of Ouesso and its market served as a crucial enticement to get the Voyager, as I imagine him, on his way. That's where the wildcat notion of his wildcat journey began: Ouesso. He hadn't intended to go farther. A trip down to Ouesso and back (he *had* meant to come back, though life unfolded otherwise) would be ambitious and risky enough. But even before the idea of Ouesso, there was the dizzying happenstance of the tusks. If it was Ouesso that pulled him, it was the tusks that pushed him.

He had never gone looking for ivory. It came by accident. One day he was upriver on the Ngoko, working his net at the mouth of a feeder stream that drained from the Congo side. It was dry season—near the end of the long dry season, early March. The river was low and slow and warm, which was why he had thought the freshening flow of the feeder stream might attract fish. As it happened, not many. The catch there scarcely repaid his effort. So in midafternoon he decided to walk inland, back-following this little stream into the forest, looking for pools where small fish might be trapped and vulnerable. He fought his way along the mud banks for almost half a mile, through the thorn vines, over the cobble of roots, finding few pools and no fish. It was frustrating but not surprising. He paused for breath, dipped up a handful of water to drink, and frowned ahead, deciding whether to continue. That's when he noticed a large gray mound in the stream bottom about forty yards on. To you or to me it would have looked like a granite boulder. But there are no granite boulders in northern Congo or southeastern Cameroon, and the Voyager had never seen one. He knew immediately what it was: an elephant. His heartbeat surged and his first instinct was to run.

Instead he stared. His legs didn't go. He lingered, unsure why. He sensed terror in the scene somewhere, but the terror wasn't his. Then he realized what seemed wrong—the elephant was down, and not in a position that might suggest sleep. Its face lay smashed into the mud, its trunk sideways, its hip canted up. He approached carefully. He noticed the purplish red holes along its lower sides and belly. Protruding from one of those holes was a Baka spear. He could see the awful way the beast had collapsed down over its left shoulder, its front leg on that side bent out at a ruinous angle. By the time he had crept within ten yards, he knew that it was dead.

A sizable male, a youngish adult, with good ivory. Left to die alone in a stream bottom and rot. Quickly the Voyager made some deductions. Probably it had been killed by a hunting party of Baka men—but not quite killed, just mortally wounded. It had broken away, escaped, and to do that, presumably, it would have had to kill one or two of the Baka who surrounded it. The others must have lost heart for the chase. Maybe this had occurred on the north side of the river. Maybe the elephant, wounded and desperate, had swum across. But if the Baka took up the trail, got themselves over here, and reappeared now—that could be bad for him. Finding the Voyager with their costly trophy, the Pygmies might fill *him* with purple spear holes. Assuming they could summon the audacity to murder a Bantu. So he worked fast. He whaled into the elephant's face with his machete, hacking through flesh and gristle, opening an ugly maw that no longer looked elephantine but like something else, something exploded and ogrish, and within half an hour he had twisted both tusks free. They surrendered with ripping noises, like any tooth drawn from its jaw.

He shaved the tusks free of tissue, then rubbed them with sandy mud and rinsed them white in the stream. Held in his hands, each one seemed huge. Bounteous. Maybe fifteen kilos. He had never experienced such an armload of wealth. He could only handle one at a time. He examined each in turn, passing his hand down the smooth white curve to the point. Then he gathered up both and staggered back to his canoe, crouching and dodging through the vines, and dropped them into the bilge with his few fish. Untied the boat quickly, caught the current, headed downstream. Having rounded one bend, he began to ease, his heart slowing back to normal.

What had just happened? He had stumbled upon half a fortune and stolen it, that's what. Claimed it, rather. Now what?

Back at his camp, the Voyager cached the tusks hastily beneath leaves and branches in a recess beside a fallen tree. Midway through the first night he woke, suddenly aware that his hiding place was inadequate, stupidly so, and he waited out the darkness impatiently. At daylight he rose, scraped away the coals and embers and ash from his campfire—his hearth site of several years' custom—and dug a pit on that spot, cracking through the layer of baked earth with his machete, slapping deep slices into the clay beneath. He went down four feet. He shaped a deep, narrow slot. He wrapped the two tusks in *ngoungou* leaves for protection and nestled them at the bottom of the trench. Then he refilled it, leveled the ground carefully, spread the old ashes back where they'd been, replaced the charred logs, and lit a new fire. Now his treasure was safe, maybe, for a while. And he could think about what to do.

There were no easy answers. There was opportunity and there was risk. He was not a man who hunted elephant, and everyone who knew him knew that. He was not supposed to have tusks. If he took them to Moloundou the agents of the French concessionaires, greedy for ivory, leaching it from the forest by all manner of compulsion and threat, would simply impound them. He might even be punished. Others would try to steal them, or to trade for them while cheating him of their value. He thought through the scenarios. He wasn't a cunning man but he was tough and stubborn.

Six months passed. He continued to live as before: fishing the river, drying fish at his camp, spending his days alone, making infrequent stops at Ngbala or Moloundou for trading. There was one man in Moloundou, a merchant, not a local Bantu and not a concessionaire's agent but a half-Portuguese outsider with connections, notoriously clever, known to deal discreetly in elephant meat and ivory. One day during a transaction over fish, salt, and fufu, the Voyager asked this merchant about the price for tusks. *It was just a question!* The merchant looked at him slyly and mentioned a number. The number seemed high but not very high, and the Voyager's face may have flickered with disappointment. He said nothing more.

Two nights later, the Voyager returned from upriver and found his camp wrecked. The half-Portuguese merchant had spoken with someone, and that someone had gone straight to rob him.

His hut had been ripped apart, his drying racks broken. His few possessions—his second net, some tin pots, a camp knife, a shirt, his raffia mat, and the rest—had been scattered disdainfully. His little tin box had been broken open and the fishhooks and tobacco dumped out. Dried fish lay on the ground, willfully trodden upon. There were signs of digging here and there—beside the fallen log, in the floor of his hut, a couple other places too. Desultory, petulant searching. The Voyager's campfire had been scattered, logs and ashes kicked away. His breath caught when he saw that. But the dirt beneath the ashes hadn't been disturbed. They hadn't found what they had come for.

So he turned his mind toward Ouesso. He waited out the night in his ruined camp, beside a fire burning low, with his machete in hand. At dawn he excavated his tusks and, leaving them leaf-wrapped and dirty, without pausing to savor their cool precious weight, put them into his canoe. He covered the tusks with dried fish, of which he had plenty, and smoked fish, of which he had just a bit, then covered the fish with more ngoungou leaves in neat bundles, as though he were taking them to market. Ngoungou leaves had their value as wrapping, but it was minimal; a pathetic, countryman's product, and therefore plausible. Atop the leaves he placed his mat. He pushed off, paddled out, and let himself be swung downriver on the Ngoko, putting Moloundou behind him. He paddled steadily for hours, reached the Sangha, turned downstream there, and continued straight to Ouesso.

Half a mile below the town, he found an eddy and pulled his boat up into the forest. There was no landing, no trail, no camp, no sign of human presence—which was good. Next day he concealed the canoe beneath leafy branches and bushwhacked northwest until he struck the outer lanes of Ouesso. He walked straight to the market by following other people. He had never seen such a concentration of humans and, once he was amid the crowd, his heart began thumping as it had when he stood over the dead elephant. But no one hurt him; no one even looked at him twice, despite the fact that his clothes were shabby and he carried a machete. He saw other men in dirty clothes, a few, and one or two of them carried machetes also. He began to relax.

The market, sheltered in a huge round building with a metal roof, was wondrous. You could buy meat, you could buy fish, you could buy colorful clothing and dried manioc and vegetables and fishnets and things he had never seen. The Voyager had no money of any sort, not francs, not brass rods, but he wandered among the goods as though he might want something. He admired the duikers and the monkeys. He picked up a gorilla hand, while the seller woman watched him closely, and set it back down. The people spoke Lingala. He exchanged a few words with a man selling fish. The Voyager was more cautious than he had been in Moloundou. Do you buy smoked fish if I have some? he asked. Maybe, when I see it, the man said. The Voyager took note of another man nearby, behind a plank table upon which sat large chunks of elephant meat, smoky and gray. A man who sold elephant meat might also deal in ivory. The Voyager memorized that man's face but didn't speak with him. He would do it tomorrow.

He walked back out of town and into the forest, satisfied by his judicious preliminary excursion, and when he emerged through the undergrowth to his riverbank hiding spot, he was horrified to see the cut branches cast aside and someone bent over his boat. Horrified and enraged: at himself for his redoubled stupidity, at the world, and especially at the man coveting his tusks. The Voyager raised his machete, ran forward, and struck before the interloper had half turned around, splitting the man's skull like a dry coconut. That made a sickening, fateful sound. The man fell hard. Where his head had broken open, pink brains showed and blood surged around the pinkness, then stopped.

It was scarcely midafternoon of the Voyager's first day in Ouesso and he had killed someone. What sort of hellish place was this?

His next shock came when he rolled the dead man over. It wasn't a man's face; it was a boy's. Smooth skin, baby cheeks, long jaw, barely old enough for initiation. The Voyager had been fooled by height. He had killed a tall youngster, a gangly boy who had dared to stoop over his canoe. A boy from the town, with relatives who would miss him. This wasn't good.

The Voyager stood for a moment, exhausted and pained, calculating his situation. Then again he moved quickly. He dragged the boy's body to the river. Splashing into the shallows, stumbling, he pulled it offshore just enough to be sure of current, released it, and watched it drift away. The body floated low in the water but it floated. Back on the bank, he rifled down into his canoe and confirmed that the tusks were still there. They were. He gripped each individually at the tip, assuring himself: one, two. He peeled back the leaf wrap and looked. Yes, ivory, two tusks. He dragged his canoe to the water, climbed in, and began paddling downstream. Within fifty yards he caught up with the boy's body and passed it by. He did not glance back toward Ouesso.

Now he was launched, untethered, no going back. For three weeks he journeyed downstream. Or maybe four weeks; he didn't keep a tally of the days. He had his canoe and his tusks, his machete, his fishing line and hooks, little else. His immediate purpose was to stay alive, day by day. His driving goal was to recoup a life from the ivory. He resumed fishing as he went, trolling with his line, seldom stopping except for the night. He ate what he caught, saving the dried and smoked fish for contingency. He was on the water again every morning by full light. He passed another town, avoiding it along the far bank, and paddled through a stretch where the river meandered slowly amid swamplands. He could see it was taking him generally south. There were adventures and mishaps and some further narrow escapes along the way. Maybe you can imagine them as well as I can. There was the encounter with the men on the log raft, drifting downriver, to whom he sold fish and by whom he was warned about the Bobangi, an imperious people controlling trade and passage at the mouth of the Sangha. He didn't know what that meant, the mouth of the Sangha; he pictured this river going on forever. There was the ambush by the crocodile, another hateful moment, but he had been lucky that morning. It was a nasty animal, not large, barely six feet, presumptuous and stupid to attack a human, and he'd had his revenge. He ate the belly meat and tail of the crocodile for six days afterward. He had never eaten chicken so to him it tasted like fish. He placed the crocodile's severed head into a column of driver ants and they cleaned it of flesh within an afternoon. Now the sun-bleached skull rode atop the other cargo in his canoe, toothy and grinning, like a totem. He reached the mouth of the Sangha and tried to elude the Bobangi, running midriver at night and laying up by day. But he couldn't stay with his treasures every moment. He left the boat unguarded once, for only a short time, to gather fruit beneath a *mobei* tree, and so there was his standoff with the solitary Bobangi man whom he found, as he had found the Tall Boy, committing an outrage: looking into his canoe. Unlike the Tall Boy, this man heard him and turned around.

The man had gray hair at his temples and his left eye was milky blue. His right eye was normal. He was old but not too old to be dangerous; his body appeared still strong. He carried a small iron knife, but no machete, and a little packet in animal hide strung around his neck. He looked like a magus or a sorcerer. He had unwrapped the Voyager's ivory. The Voyager knew that there were many other Bobangi on the river, maybe even some within earshot. The Voyager felt trapped. He remembered the sickening sound of his machete on the Tall Boy's head. He decided, very quickly, upon a desperate compromise. He addressed the blue-eyed man in Lingala, not sure whether a Bobangi would understand.

I give you one tusk, the Voyager said.

No sign of response.

I give you one tusk, he repeated, speaking very clearly. *You deliver it to your chief. Or . . . you don't.*

He waited, letting the blue-eyed man ponder.

One tusk, he said. He held up a finger. *Or I fight you and I kill you for two.*

It seemed a long delay. The Voyager began wishing he had simply cracked the man's skull, at least tried to, whatever the consequences. Then the blue-eyed man turned back to the Voyager's canoe. He rummaged, shoved away leaves, and, with an effort that showed in his back muscles, lifted out one tusk. He stroked it, testing the smooth cool surface, and appeared satisfied. The Voyager watched him; willed him on his way. *All right. Take it. Go.* But then, no, the man stooped again. He picked up a single smoked fish. He gaped back at the Voyager with an expression of shameless, bemused defiance. The blue eye twitched—or was that a wink? He took the tusk and the fish and he departed.

That night the Voyager passed onward through Bobangi territory, slipping by their big village near the mouth of the Sangha, where this river debouched into another, unimaginably huge: the Congo. He was astonished when daylight revealed the extent of its braiding channels, islands, and strong currents. It was like a bundle of rivers, not just one. He paddled harder than ever now, but also more carefully, learning wariness of the eddy lines that could knock a canoe sideways, the whirlpools that could suck it under. He kept a distance between himself and other canoes. When he saw men on a raft, he paddled within shouting distance, offered to sell fish, sought information. Once he encountered a steamboat, like a great house proceeding upriver under power, with a machine inside thumping stupidly, passengers and bundled cargo on the deck. It was a strange sight. But the Voyager had seen other strange sights—the spilled brains of a boy, the Ouesso market, a blue-eyed Bobangi thief—and by now felt almost inured to astonishment. The boatman on this noisy, belching vessel, he could see, was a white man. The Voyager hugged the opposite shore.

The river continued south. He entered the territory of the Tio, a more tractable people than the Bobangi—eager for trade but not demanding monopoly, according to what the Voyager heard. Maybe the Tio were humbler because the river was now so vast. No one could imagine himself owning such a river. No tribe, even. Here the Voyager saw dozens of other boats. It was a new universe. Many canoes, several more steamboats, people hollering and trading from one boat to another. The maze of channels and the traffic, plus the increasing distance from Ouesso, gave a sense of jumble and anonymity and security that allowed the Voyager to travel by daylight, which was fortunate in these formidable waters. He sold fresh fish to Tio boatmen and swapped fish for manioc. He chatted. *Yes, I've come from the upper river, very far.* But he didn't say which river. He didn't mention ivory. He gathered intelligence without revealing much. He was tired.

He had an intermediate goal now, between the daily purpose of survival and the dream of due reward for his troubles. He had a destination: a place called Brazzaville. It was a large town, downriver, some days ahead. It sat on the right, beside a great pool. He would know it when he saw it—so he'd been told. Another big town sat on the left bank, across the pool, but that one was owned by the Belgians. *Who are the Belgians?* he asked. *Are they a tribe like the Bobangi?* Worse. Yes, he heard, Brazzaville was a good market for fish or whatever you had.

And so the Voyager arrived. He rounded a last bend, came to a great pool where the river seemed as wide as it was long, put a large island to his left as advised, and saw white buildings on the right bank, some of them twice as tall as a house, taller even than the circular market hall at Ouesso. He paddled toward the white buildings. Coming near, he held himself some distance out, drifting, observing, until he was well past the docks and the big boats and the bustle of workingmen, then beached his canoe in a quieter place. Several children gaped, as children do, but no one else noticed him. People were busy and no adults diverted their attention to the sight of a strong young Bakwele coming ashore in tattered clothes with a crocodile skull, a single fine tusk, and half a boatload of rotten fish.

He stepped out of the water and stood alone. No one greeted him.

No one knew what he had done. No one compared him to Lewis and Clark. No one hailed him as the Marco Polo of the upper Congo basin. No one knew that he was Huck Finn and Jim, John Wesley Powell on the Colorado, Teddy Roosevelt on the River of Doubt, Frank Borman circling the moon in *Apollo 8*, and Dr. Richard Kimble at large. No one knew.

The Voyager walked into town and sold his tusk the first afternoon, receiving 120 brass rods, which was a good price, he thought, but also somehow anticlimactic and unsatisfactory. For his crocodile skull, at the benign whim of the ivory buyer, he received another ten brass rods. He bought some palm wine, got drunk, found that experience not to his liking, and never did it again. The rest of his money he saved, or rather set aside, spending it slowly and variously until it was gone. He had arrived.

He found lodging in Poto-Poto, a neighborhood east of the city center, full of others from the upper river, and got work on the waterfront. He made friends. He settled in. Urban life suited him. He became something of a colorful figure, confident, charming in his river-man way, with stories to tell. No one viewed him as the pariah son of a sorceress. No one guessed that he had ever been a surly young loner. No one knew his real name because he had invented another. And the other thing no one knew, not even he, was that he had brought a new element, a new circumstance, to Brazzaville. A virus, in his blood. More specifically: He had brought HIV-1 group M.

Seven and eight and nine years later, near the end of his life, the Voyager would tell some of his stories to friends, acquaintances, and a few of the women with whom he had relationships, transient or longer: about the Dead Elephant, the Half-Portuguese Merchant, the Tall Boy, the Crocodile, and the Blue-Eyed Bobangi. In his telling, the Tall Boy became an adult and the Crocodile was very large, a leviathan. No one doubted his word. They knew he had come down the river and it must have been perilous. The crocodile skull wasn't there to belie him. During those years he slept with thirteen women, all of whom were femmes libres to one degree or another. One of those, a young

Tio girl who had recently arrived in Brazzaville from upriver, and who found that she fancied him more than she did her freedom, became his wife. Eventually he infected her with the virus. He also infected one other, a rather more professional woman who lived in a small house in the Bacongo neighborhood, west of town, where he visited her on an intermittent basis when his wife was pregnant. The other eleven women had only fleeting sexual contacts with him and were luckier. They remained HIV-negative. The Voyager's personal lifetime contribution to the basic reproduction rate was therefore precisely 2.0. People liked him and were sorry when he fell ill.

The Bacongo girlfriend was vivacious and pretty and ambitious for wider horizons, so she crossed the pool to Léopoldville, where she had a successful career, though not a long one.

18

If the virus reached Léopoldville in 1920 or so, that still leaves a gap of four decades to the time of ZR59 and DRC60, those earliest archival HIV sequences. What happened during the interim? We don't know, but available evidence allows a rough sketch of the outlines of possibility.

The virus lurked in the city. It replicated within individuals. It passed from one person to another by sexual contact, and possibly also by the reuse of needles and syringes for treatment of well-known diseases such as trypanosomiasis. (More on that possibility, below.) Whatever its means of transmission, presumably HIV caused immunodeficiency, eventually death, among most or all people infected—except those who died early from other causes. But it didn't yet assert itself conspicuously enough to be recognized as a distinct new phenomenon.

It may also have proliferated slowly in Brazzaville, across the pool, helped along there too by changing sexual mores and programs of therapeutic injection. It may have lingered in villages of southeastern Cameroon or elsewhere in the upper Sangha basin. And wherever it was, but definitely in Léopoldville, it continued to mutate. The wide divergence between ZR59 and DRC60 tells us that. It continued to evolve.

Studying the evolutionary history of HIV-1 is (notwithstanding that ill-advised comment by the WHO's David Heymann to Tom Curtis) more than an idle exercise. The point is to understand how one strain of the virus, group M, has made itself so deadly and widespread among humans. Such understanding, in turn, may lead toward better measures to control the devastation of AIDS, possibly by way of a vaccine, more likely by way of improved treatments. That's why scientists such as Beatrice Hahn, Michael Worobey, and their colleagues explore the molecular phylogenetics of HIV-1, HIV-2, and the various SIVs. One issue they address is whether the virus became virulent before, or only after, its spillover from *Pan troglodytes*. To state the question more plainly: Does SIV_{cpz} kill chimps, or is it only an innocuous passenger? Answering that one could reveal something important about how human bodies respond to HIV-1.

For a while after the discovery of SIV_{cpz}, the prevailing impression was that it's harmless in chimpanzees, an ancient infection that may once have caused symptoms but no longer does. This impression was bolstered by the fact that, in the earlier years of AIDS research, more than a hundred captive chimpanzees had been experimentally infected with HIV-1 and none had shown immune system failure. When a single lab chimpanzee did progress to AIDS (ten years after experimental infection with three different strains of HIV-1), its case was remarkable enough to merit a six-page paper in the *Journal of Virology*. The researchers implied that this was good news, finally offering hope that chimpanzees do represent a relevant experimental model (that is, a sufficiently analogous test subject) for studying human AIDS. There was even a report, based on genetic analysis of captive animals in the Netherlands, suggesting that chimpanzees had "survived their own AIDS-like pandemic" more than 2 million years ago. They had emerged from the experience, according to this line of thought, with genetic adaptations that render them resistant to the effects of the virus. They still carry it but apparently don't get sick. That notion, to repeat, was founded on captive chimpanzees. As for SIV-positive chimps in the wild: No one knew whether they suffer immunodeficiency. It was a difficult question to research.

These suppositions and guesses jibed with available information about other variants of the virus in other primates. SIV is highly diverse and broadly distributed, found as a naturally occurring infection in more than forty different species of African monkey and ape. (But it seems to be unique to that continent. Although some Asian primates have acquired the virus in captivity, it hasn't shown up among wild monkeys in either Asia or South America.) Most of those SIV-carrying African simians are monkeys. Each kind of monkey harbors its own distinct type of SIV, such as SIV_{gsn} in the greater spot-nosed monkey, SIV_{ver} in the vervet, SIV_{rcm} in the red-capped mangabey, and so forth. Based on evidence presently available, none of those SIVs seems to cause immunodeficiency in its natural host. A close evolutionary kinship between two kinds of simian, such as L'Hoest's monkey and the sun-tailed monkey, both classified in the genus *Cercopithecus*, is sometimes paralleled by a close similarity between their respective SIVs. Those deep taxonomic alignments, plus the absence of noticeable disease, led researchers to suspect that African monkeys have carried their SIV infections for a very long time—probably millions of years. That length of time would allow divergence among the viruses and mutual accommodation between each type of virus and its host.

The same two-part hypothesis applied also to chimps: that their virus, SIV_{cpz}, is (1) an ancient infection that now (2) causes no harm. But for chimps those were just tenuous assumptions. Then new evidence and analyses addressed them, and both parts turned out to be wrong.

The first premise, that SIV_{cpz} has lurked within chimpanzees for a very long time, began to look doubtful in 2003. That's when another team of researchers (led by Paul Sharp and Elizabeth Bailes of the University of Nottingham, and including again both Beatrice Hahn and Martine Peeters) noticed that SIV_{cpz} seems to be a hybrid virus. The Nottingham group reached that conclusion by comparing the genome of SIV_{cpz} with the genomes of several monkey SIVs. They found that one major section of the chimp virus's genome matches closely to a section of SIV_{rcm}. Another major section closely resembles a section in SIV_{gsn}. In plain words: The chimp virus contains genetic material from the virus of red-capped mangabeys and also genetic material from the virus of greater spot-nosed monkeys. How did it happen? By recombination—that is, genetic mixing. A chimpanzee infected with both monkey viruses must have served as the mixing bowl in which two viruses traded genes. And when did it happen? Possibly just hundreds of years ago, rather than thousands or tens of thousands.

How did a single chimpanzee become infected with two monkey viruses? Presumably that occurred through predation, or through the combined circumstances of predation (bringing aboard one virus) plus sexual transmission (bringing aboard a second), followed by a chance rearrangement of genes between one virus and the other during viral replication. Chimpanzees are omnivores who love an occasional taste of meat. They kill monkeys, rip them apart, fight over the pieces or share out gobbets and joints; then they eat the flesh, red and raw. It doesn't happen often, just whenever the opportunity and the hankering arise. Such gore fests must sometimes involve blood-to-blood contact. Chimpanzees, even without the use of machetes, suffer wounds on their hands and in their mouths. Bloody meat plus an open sore equals exposure. What the Nottingham group suggested was another chimpanzee version of the cut-hunter hypothesis—except in this case the cut hunter was the chimp.

19

So the very existence of SIV_{cpz} is relatively recent. It has no ancient association with chimpanzees. And now, based on a study published in 2009, part two of the two-part hypothesis has also been cast into doubt. The virus is not harmless in its chimpanzee host. Evidence from the chimps of Gombe—Jane Goodall's study population, known and beloved around the world—suggests that SIV_{cpz} causes simian AIDS.

I've mentioned already that the first wild chimp to test SIV-positive was at Gombe. What I didn't say, but will here, is that SIV-positive status among Gombe's chimpanzees correlates strongly with failing health and early death. Again it was Beatrice Hahn and her group who made the discovery.

Having found SIV_{cpz} in captive chimps, Hahn wanted to look for it in the wild. But she and her team of young molecular biologists knew little about sampling chimpanzees in an African forest. What do you do—go out and dart one? Knock the ape out with ketamine, take blood, wake him up, and send him on his way? Egads, no! said field primatologists, horrified at the prospect of any such invasive violation of their sensitive, trusting subjects. It was a new realm for Hahn, with a new set of concerns and methods, to which she quickly became attuned. At a scientific meeting that brought primate researchers together with virologists, she met Richard Wrangham, of Harvard, highly respected for his work on the behavioral ecology and evolution of apes. Wrangham has for many years led a study of chimpanzees at Kibale National Park, in western Uganda; before that, four decades ago, he did his own PhD fieldwork at Gombe. He responded enthusiastically to Hahn's idea of screening wild chimps, and ultimately it was Wrangham, she recalled, "who convinced Jane that we were okay to work with." But before any such work began at Gombe, they looked at the chimps of Kibale, Wrangham's own research site. Crucial help came from a Wrangham grad student named Martin Muller, who in 1998 had collected urine samples for a study of testosterone, aggression, and stress. Mario Santiago, of Hahn's lab, cooked up the requisite tools for detecting SIV_{cpz} antibodies in a few milliliters of piss, and Martin Muller supplied some frozen samples from his collections at Kibale. For this part of the story, I went to Albuquerque and talked with Muller, now an associate professor of anthropology at the University of New Mexico.

The Kibale samples all tested negative for SIV. "We were slightly disappointed," Muller recalled. "That was because, at the time, the conventional wisdom was that this didn't have any negative impacts on chimps." Disappointment at negative results was allowable, he meant, because positive results would not have spelled doom for those Kibale chimps—or so it was thought. Meanwhile, though, he was getting some interesting results in his hormone study and wanted to broaden his data. He and Wrangham agreed that it might be instructive to sample a few other chimp populations for comparison. That led Muller down to Gombe, in August 2000, with his urine-collecting bottles and all the cumbersome equipment necessary to keep samples frozen. He stayed only a couple weeks, training Tanzanian field assistants to continue the collecting, and brought away just a few samples himself. Back home in the United States, he e-mailed Hahn to ask whether she would like six tubes of frozen Gombe urine, to which she replied: "YES, YES, YES." He sent them with coded labels, standard procedure, so Hahn had no way of knowing whose was whose. Two of the six tested positive for SIV antibodies. Breaking the code, Muller informed her that both samples came from a chimp named Gimble, a twenty-three-year-old male.

Gimble was a well-known member of one of the famed Gombe families; his mother had been Melissa, a successful matriarch, and his brothers included Goblin, who rose to be the community's alpha male and lived to age forty. Gimble's life and career would be different—and shorter.

Soon after getting the results on Gimble, Beatrice Hahn wrote a long e-mail to Jane Goodall, explaining the context and the implications. Goodall herself had trained as an ethologist (she took a PhD at Cambridge), not as a molecular biologist, and the realm of Western blot analysis for antibodies was as alien to her as field sampling had been to Hahn. Her work on chimpanzees began back in July 1960, at what was then the Gombe Stream Game Reserve, on the east shore of Lake Tanganyika, and which later became Gombe National Park. She established the Gombe

Stream Research Center in 1965, based in a small concrete building near the lake, and continued her study of chimps in the hilly forest for another twenty-one years. In 1986 Goodall published an imposing scientific opus, *The Chimpanzees of Gombe*, and then ended her own career as a field scientist because, appalled by the treatment of chimpanzees in medical labs and other captive situations around the world, she felt obliged to become an activist. The study of Gombe's chimps went ahead in her absence, thanks to well-trained Tanzanian field assistants and later generations of scientists, adding decades of data and precious continuity to what Goodall had started. She remained closely connected to Gombe and its chimps, both personally and through the programs of her Jane Goodall Institute, but she wasn't often present at the old research camp, apart from stolen interludes of rest and reinvigoration. Instead she traveled the world, roughly three hundred days a year, lecturing, lobbying, meeting with media people and schoolchildren, delivering her inspirational message. Hahn understood the intensity of Goodall's protectiveness toward chimps in general, toward Gombe's chimps in particular, and of her wariness toward anything that might put them in more jeopardy of exploitation, especially in the name of medical science. At the end of the long e-mail, Hahn wrote:

> Let me finish by saying that finding SIV_{cpz} in the Gombe community is a virologist's DREAM-COME-TRUE. Given the wealth of behavioural and observational data that you and your colleagues have collected over decades, it is the IDEAL setting to study the natural history, transmission patterns and pathogenicity (or lack thereof) of natural SIV_{cpz} infection in wild chimpanzees. Moreover, all this can be done entirely non-invasively. AND there certainly are funding opportunities for such a unique study. So, the virologist's dream-come-true does

> not have to be the primatologist's nightmare, although I am sure it will take some time before I can convince you of that.

Eventually she did convince Goodall, but not before another nightmarish discovery emerged from the work.

Earlier in her e-mail, Hahn had written: "With respect to the chimpanzees, it is probably safe to say that SIV infection will NOT cause them to develop immunodeficiency or AIDS." On that point, she would prove herself wrong.

20

Jane Goodall described her own concerns when I caught up with her during one of her stopovers. We knew each other from previous adventures—among chimps in the Congo basin, among black-footed ferrets in South Dakota, over single-malt scotch in Montana—but this was a chance to sit down quietly at a hotel in Arlington, Virginia, during a paralyzing snowstorm, and talk about Gombe. The fiftieth anniversary of her own chimp study was approaching, and I was assigned by *National Geographic* to write about it. After we discussed her childhood influences, her dream of becoming a naturalist in Africa, her mentor Louis Leakey, her early days in the field, and her time as a PhD student at Cambridge, she herself mentioned genetics and virology. At that point I turned the conversation to SIV.

"I was really, really apprehensive about Beatrice Hahn's research," Jane volunteered. "We were, a lot of us, really nervous about the result of what might happen if she found HIV/AIDS."

She had met Hahn, talked with her, and was reassured by the force of Hahn's concern for the chimps' welfare. "But still. I still have this unease because, even though she cares, once these results are out, as they are now, other people can use them in different ways." For instance? What sort of dangers, I asked, did Jane have in mind? "That this would start a whole new flurry of research on captive chimps in medical labs." The news of chimps with AIDS, she feared, would sound like a promising opportunity to learn more about AIDS in humans, never mind the chimps.

What about the impact of the virus at Gombe itself? We both knew that Hahn had found something resembling AIDS, and by now Gimble was dead. What about the prospect that other members of the Gombe community might die of immune failure? "Yeah, exactly," Jane said. "That's a very scary thought."

As scary as it was, though, she had realized from the start of her conversations with Hahn that such a finding could be taken two ways. On the one hand, she said, there was a possible consolation: If people heard that wild chimps carry an AIDS-causing virus, they might stop hunting and butchering and eating them. "Because they'll be afraid. That was one side of it. Then the other side of it was, well, people will say, 'All these creatures are really dangerous for us, so let's kill them all.' It could have gone either way." Jane is a perspicacious woman. She has the aura of a secular saint but is actually quite human, grounded, savvy, and capable of ambivalence. As things have transpired so far, she noted, neither of the extreme outcomes has occurred.

Briefly we discussed Hahn's noninvasive sampling methodology: Urine might contain antibodies, and feces could yield viral RNA. Jane allowed that it had been reassuring, that part about no necessity of knocking chimps out and jabbing them with needles. "Don't need blood," she said. "Just need a bit of poo." Amazing what they can do from a bit of poo, I agreed.

So she had given her consent for Hahn's study, and the work proceeded. At the end of November 2000, Hahn's lab in Alabama received the first batch of material, which included three fecal samples from poor Gimble. Hahn's grad student Mario Santiago did the screening, and again all three of Gimble's samples tested positive. Santiago then amplified a viral RNA fragment and sequenced it, confirming that Gimble's virus was indeed SIV_{cpz}. It seemed to be a new strain, distinct enough from other known strains that it might be unique to East Africa. This was significant on several counts. Yes, the chimps of Gombe were infected. No, they couldn't be source animals for the human pandemic. The variants of SIV found by Martine Peeters in western Africa (this was before Hahn's own findings from Cameroon) more closely matched HIV-1 group M than the Gombe virus did.

In mid-December, another e-mail from Hahn's computer went out to Richard Wrangham, Jane Goodall, Martin Muller, and others. Under the subject line GOOD NEWS AT LAST, Hahn described the findings from Gimble and the position of his strain on the SIV family tree. Then, with her characteristic penchant for uppercase exuberance, she wrote: "THIS IS A HOME RUN!"

21

That was just the beginning. For nine years the study continued. Fieldworkers at Gombe collected fecal samples from ninety-four different chimpanzees, each of which was known by name and, in most cases, by its individual character and family history. Beatrice Hahn's people did the analyses, finding that seventeen of those ninety-four chimps were SIV-positive. As time passed, some chimps died. Others disappeared in the forest and were presumed dead when they failed to reappear. Death is often a private matter for wild creatures, including chimpanzees, especially when it comes upon them by slow and painful degrees. They tend to go absent from the social group, if there is a social group, and meet the end alone. Gimble last showed himself to trackers on January 23, 2007. His body was never found.

Back in Birmingham, there was turnover of a different sort, as grad students and postdocs cycled through Hahn's lab. Mario Santiago departed, heading off for the next stage of his career, and Brandon Keele arrived. Samples continued coming from Gombe, in occasional batches, and those samples were analyzed—a slow and laborious process. Much of the work fell to Keele, though even for him this was "a backburner project." Keele described to me, during my visit with him at Fort Detrick, the moment of recognition that occurred near the end of his postdoc period, bringing that project to the front burner.

"I was trying to leave and finish up. I said to myself, 'I wonder what's happening with these chimps?'" He was aware that the number of known SIV-positives at Gombe had increased as the sampling stretched on, and that there was evidence of vertical transmission (mother to infant) as well as sexual transmission accounting for new infections. He thought the study might yield an interesting, undramatic paper about how a harmless virus spreads through a population. "And then we started compiling the data," he told me. That meant bringing in a dimension of behavioral observations from the field. So he contacted collaborators at the Jane Goodall Institute's research headquarters in

Minnesota and, asking about one individual after another, heard a drumbeat of unsettling news.

"Oh, no, that chimp is dead."

"No, that chimp is dead. He died in 2006."

"No, that chimp is dead."

Keele recalled asking himself: "What the hell is going on?" Part of the answer, revealed when he saw an updated mortality list, was that a wave of untimely deaths had been sweeping through SIV-positive members of the Gombe population.

He and the team at Hahn's lab had recently written an abstract for a talk he planned to give at a meeting, which would lead in time to a journal publication. The draft abstract, by Keele's recollection, contained a sentence such as: "It doesn't really seem that there is a death hazard to infection in these chimps." They had sent the draft to their partners at Gombe, who responded quickly with news of seven additional chimpanzee deaths, about which Keele hadn't even known. He scrapped the abstract, thought again about what he was doing, and began working more closely with Gombe and Minnesota to assemble a more complete set of data. Then they would see where it led.

Around the same time, spring of 2008, Keele also heard about some unusual pathology results on tissues from one dead Gombe chimp. The chimp was known as Yolanda, a twenty-four-year-old female. She sickened in November 2007, of an unknown ailment, and came down from the mountains to languish near the research center. People tried to feed her, but Yolanda didn't eat. She sat in the rain amid thick vegetation, weakened and miserable, and then died. They put her body in a freezer. Two months later, it was thawed for necropsy.

The necropsy was performed by Jane Raphael, a Tanzanian veterinarian working at the Gombe Stream Research Center and specially trained for the task. Not knowing whether Yolanda had been SIV-positive, Raphael took the stipulated precautions. She wore a full Tyvek suit, two layers of gloves, an N95 respirator mask, a face shield, and rubber boots. She split open Yolanda's belly, cut through the ribs, and spread them wide to see what she could see.

"The main problem was in the abdominal cavity," Raphael told me, two years later, as we sat in her small office just up from the shore of Lake Tanganyika. "There was something like abdominal peritonitis. The intestines were very much adhered together." Raphael, a quiet woman, wearing a neat cornrow hairdo and a flowered print dress, chose her words carefully. She described separating the glommed guts with her gloved hands. "It was unusual," she said. She seemed to remember it all vividly. "The muscles underneath the pelvis were very much inflamed. Red. And they had some blackish spots." What caused the inflammation? Judiciously empirical, Raphael said she didn't know.

Her inspection done, she had snipped out tissue samples from virtually every organ: spleen, liver, intestines, heart, lungs, kidneys, brain, lymph nodes. For the SIV-positive cases, she said, lymph nodes were especially important. Yolanda's lymph nodes appeared normal to the eye, but histopathology would later penetrate that illusion. Some of the samples, preserved in RNAlater, went off to Beatrice Hahn. Others, pickled in formalin, were destined for a pathologist in Chicago. When the results came together, this case would challenge prevailing ideas about SIV in chimpanzees. "Previously it was said, they are infected but they don't come down with the disease," Raphael told me. "Yolanda made us to start thinking otherwise."

I followed the pickled samples to Chicago, where the pathologist who had examined them, Karen Terio, welcomed me to a glimpse of the evidence. Terio trained as a veterinarian, at one of the country's best vet schools, then did a residency and a doctorate in pathology, specializing in diseases transferred between species. She worked for the University of Illinois and consulted for the Lincoln Park Zoo, which helps run a health-monitoring project at Gombe. Hence the lymph nodes and other bits of Yolanda came for her expert scrutiny. Terio cut up the tissues, sent them to laboratory technicians for mounting and staining, then sat down for a look at the slides. "It was striking because I couldn't find any lymphocytes," she told me. "When I saw the first lymph node, I thought, 'Hmm, this is weird.'" She asked her boss to have a look through the microscope. He did, and agreed there was something very wrong. She phoned a colleague at the Lincoln Park Zoo, Elizabeth Lonsdorf, who leads the zoo's work on behalf of wild African apes, including the health project at Gombe.

"We have a problem," Terio told Lonsdorf. "She doesn't have any lymphocytes."

"Does that mean what I think it means?"

"Yes. The lesions in this animal look like an end-stage AIDS patient."

Together she and Lonsdorf made a call to Beatrice Hahn. Hahn's first question was, "Are you sure?" Terio was indeed sure, but she quickly e-mailed images of the slides so that the others could judge for themselves. Brandon Keele was by now in the loop. Terio sent actual slides to another collaborator, an expert on immune-system pathology, to refine the diagnosis. Everyone agreed and, with the sample code broken, everyone knew how these pieces fit together: The chimp Yolanda, dead at age twenty-four, had been SIV-positive and suffering immunodeficiency.

Inviting me to a chair at her big double-viewer Olympus microscope, Karen Terio brought out the same slides she had shared with Hahn and Lonsdorf. From her place at the scope she could manipulate a cursor, a little red arrow, moving it over the field to point out what we were seeing. First she showed me a thin-slice section from a lymph node of a normal, SIV-negative chimpanzee. This was for comparison. It looked like a peat bog as viewed on Google Earth, bulging and rife with sphagnum and huckleberry, thick, rich, and riddled just slightly with narrow spaces resembling small sloughs and creeks. The tissue was stained magenta and heavily speckled with darker blue dots. The dots, Terio explained, were lymphocytes in their healthy abundance. In an area where they're especially dense, they pack together into a follicle, like a bag full of jellybeans. She jabbed her red arrow at a follicle.

Then she placed another slide into viewing position. The slide held a slice from one of Yolanda's lymph nodes. Instead of a peat bog, it looked like scrub desert slashed by a large drywash, many days since the last rain.

"Mmmm," I said.

"This is essentially the connective tissue," Terio said. She meant that it was supportive structure only, minus the working innards. Sere and empty. "We've got very, very few lymphocytes left in this animal."

"Yeah."

"And it's collapsed. You see, this whole thing has just sort of collapsed on itself, 'cause there's nothing in there to hold it up." Her little red arrow wandered forlorn through the desert. No sphagnum, no follicles, no little blue dots. I imagined Karen Terio, back in April 2008, examining these slides on her lonesome—and encountering such evidence, before anyone else, at a time when the illusion of nonpathogenic SIV_{cpz} was embraced by researchers everywhere.

"So you sat there, and looked at this . . ."

"And went, 'Oh, no,'" she said.

22

Terio's findings, plus the field data from Gombe, plus the molecular analyses from Hahn's lab—these all came together in a paper published by *Nature* during the summer of 2009. Brandon Keele was first author; Beatrice Hahn was last. "Increased Mortality and AIDS-like Immunopathology in Wild Chimpanzees Infected with SIV_{cpz}" was the catchy title. I think of it—and I'm not alone—as "the Gombe paper." The long list of coauthors included Karen Terio, Terio's boss, Elizabeth Lonsdorf, Jane Raphael, two of Hahn's senior colleagues, the expert on primate cell pathology, the chief scientist at Gombe, and Jane Goodall herself.

"Well, I sort of had to be. But I had these long talks with Beatrice first," Jane told me. "She was going to publish it anyway." In the sweep of inevitability and the name of science, Dr. Goodall signed on.

The paper's salient conclusion was that, contrary to Keele's earlier draft abstract, there is indeed a death hazard for SIV-positive chimps at Gombe. Of the eighteen individuals that died during the study period, seven were SIV-positive. Given that less than 20 percent of the population was SIV-positive, and adjusted for normal mortality at a given age, this reflected a risk of death ten times to sixteen times higher for SIV-positive chimps than for SIV-negatives. Repeat: ten to sixteen times higher. The total numbers were small but the margin was significant. Infected animals were falling away. Furthermore, SIV-positive females had lower birth rates and greater infant mortality. Further still, three necropsied individuals (including Yolanda, though her name wasn't mentioned) showed signs of lymphocyte loss and other damage resembling end-stage AIDS.

The authors suggested, cautiously but firmly, "that SIV_{cpz} has a substantial negative impact on the health, reproduction and lifespan of chimpanzees in the wild." So it's not a harmless passenger. It's a hominoid killer. Their problem as well as ours.

23

Here's what you have come to understand. That the AIDS pandemic is traceable to a single contingent event. That this event involved a bloody interaction between one chimpanzee and one human. That it occurred in southeastern Cameroon, around the year 1908, give or take. That it led to the proliferation of one strain of virus, now known as HIV-1 group M. That this virus was probably lethal in chimpanzees before the spillover occurred, and that it was certainly lethal in humans afterward. That from southeastern Cameroon it must have traveled downriver, along the Sangha and then the Congo, to Brazzaville and Léopoldville. That from those entrepôts it spread to the world.

Spread how? Once it reached Léopoldville, the group M virus seems to have entered a vortex of circumstances unlike anything at the headwaters of the Sangha. It differed from HIV-2 biologically (having adapted to chimpanzee hosts) and it differed from groups N and O by chance and opportunity (having found itself in an urban environment). Whatever happened to it in Léopoldville during the first half of the twentieth century can only be conjectured. Population density of potential human hosts, a high ratio of males to females, sexual mores different from what prevailed in the villages, and prostitution—these were all parts of the mix. But sex plus crowding may not be a sufficient explanation. A fuller chain of conjecture, and maybe a better one, has been offered by Jacques Pepin, a Canadian professor of microbiology who, during the 1980s, worked for four years at a bush hospital in Zaire. Pepin coauthored several journal papers on the subject and, in 2011, published a book titled *The Origins of AIDS*. Having added some deep historical research to his own field experience and microbiological expertise, he proposed that the crucial factor intermediating between the Cut Hunter and the global pandemic was the hypodermic syringe.

Pepin wasn't referring to recreational drugs and the works shared by addicts at shooting galleries. In a paper titled "Noble Goals, Unforeseen Consequences," and then at greater length in his book, he pointed instead to a series of well-intended campaigns by colonial health authorities, between 1921 and 1959, aimed at treating certain tropical diseases with injectable medicines. There was a massive effort, for instance, against trypanosomiasis (sleeping sickness) in Cameroon. Trypanosomiasis is caused by a persistent little bug, *Trypanosoma brucei*, transmitted in the bite of tsetse flies. The treatment in those years entailed injections of arsenical drugs such as tryparsamide—and a patient didn't get just one shot but a series. In Gabon and Moyen-Congo (the French colonial name for what's now the Republic of the Congo), the regimen for trypanosomiasis sometimes entailed thirty-six injections over three years. And there were similar efforts to control syphilis and yaws. Malaria was treated with injectable forms of quinine. Leprosy patients, in that era before oral antibiotics, underwent a course of injections with extract of chaulmoogra (an Indian medicinal plant), two or three shots per week for a year. In the Belgian Congo, mobile teams of *injecteurs,* people with no formal education but a small bit of technical training, visited trypanosomiasis patients in their villages to give weekly shots. It was a period of mania for the latest medical wonder: needle-delivered cures. Everyone was getting jabbed.

Of course, this was long before the era of the disposable syringe. Hypodermic syringes, for injecting medicines into muscles or veins, were invented in 1848 and, until after World War I, were handmade of glass and metal by skilled craftsmen. They were expensive, delicate, and meant to be reused like any other precision medical instrument. During the 1920s their manufacture became mechanized, to the point where 2 million syringes were produced globally in 1930, making them more available but not more expendable. To the medical officers working in Central Africa at that time, they seemed invaluable but were in short supply. A famous French colonial doctor named Eugène Jamot, working just east of the upper Sangha River (in a portion of French Equatorial Africa then known as Oubangui-Chari) during 1917–1919, treated 5,347 trypanosomiasis cases using only six syringes. This sort of production-line delivery of injectable medicines didn't allow time for boiling a syringe and needle between uses. It's difficult now, based on skimpy sources and laconic testimony, to know exactly what sort of sanitary precautions were taken. But according to one Belgian doctor, writing in 1953: "The Congo contains various health institutions (maternity centres, hospitals, dispensaries, etc.) where every day local nurses give dozens, even hundreds, of injections in conditions such that sterilisation of the needle or the syringe is impossible." This man was writing about the risk of accidental transmission of hepatitis B during treatment for venereal diseases, but Pepin quoted his report at length, for its potential relevance to HIV:

> The large number of patients and the small quantity of syringes available to the nursing staff preclude sterilisation by autoclave after each use. Used syringes are simply rinsed, first with water, then with alcohol and ether, and are ready for a new patient. The same type of procedure exists in all health institutions where a small number of nurses have to provide care to a large number of patients, with very scarce supplies. The syringe is used from one patient to the next, occasionally retaining small quantities of infectious blood, which are large enough to transmit the disease.

How much of this went on? Very much. Pepin's diligent search through old colonial archives turned up some big numbers. In the period 1927–1928, Eugène Jamot's team in Cameroon performed 207,089 injections of tryparsamide, plus about a million injections of something called atoxyl, another arsenical drug for treating trypanosomiasis. During just the year 1937, throughout French Equatorial Africa, the army of doctors and nurses and semipro injectors delivered 588,086 needlesticks aimed at trypanosomiasis, not to mention countless more for other diseases. Pepin's arithmetic totaled up 3.9 million injections just against trypanosomiasis, of which 74 percent were intravenous (right into a vein, not just a muscle), the most direct method of drug delivery and also the best for unintentionally transmitting a blood-borne virus.

All those injections, according to Pepin, might account for boosting the incidence of HIV infection beyond a critical threshold. Once the reusable needles and syringes had put the virus into enough people—say, several hundred—it wouldn't come to a dead end, it wouldn't burn out, and sexual transmission could do the rest. Some experts, including Michael Worobey and Beatrice Hahn, doubt that needles were necessary in any such way to the establishment of HIV in humans—that is, to its early transmission from one person to another. But even they agree that injection campaigns could have played a role later, spreading the virus in Africa once it was established.

This needle theory didn't originate with Jacques Pepin. It dates back more than a decade to work by an earlier team of researchers, including Preston Marx of the Rockefeller University, who proposed it in 2000 at the same Royal Society meeting on AIDS origins at which Edward Hooper spoke for his oral polio vaccine theory. Marx's group even argued that serial passage of HIV through people, by means of such injection campaigns, might have accelerated the evolution of the virus and its adaptation to humans as a host, just as experimentally passaging some other virus through a series of laboratory mice might assist its adaptation to, and increase its transmissibility among, mice. Jacques Pepin picked up where Preston Marx left off, though with less emphasis on the evolutionary effect of serial passage. Pepin's main point was simply that dirty needles, used so widely, must have raised the prevalence of the virus among people in Central Africa. Unlike the OPV theory, this one hasn't been discredited by further research, and Pepin's new archival evidence suggests that it's highly plausible, if unprovable.

Most of those injections for trypanosomiasis occurred in the countryside. City dwellers were less exposed to trypanosomiasis, partly because the tsetse fly doesn't thrive in urban jungles as well as it does in green ones. One question that occurred to me, therefore, was whether any such mania for injecting had also gripped Léopoldville, where HIV met its most crucial test. Pepin's answer is unexpected, interesting, and persuasive. Never mind trypanosomiasis. He discovered a different but equally aggressive campaign of injections, aimed at limiting syphilis and gonorrhea in the city's population.

In 1929, the Congolese Red Cross established a clinic known as the Dispensaire Antivénérien, open to women and men for the treatment of what we used to call venereal diseases. Located in a neighborhood on the east side of Léopoldville, near the river, it was a private facility providing a public service. Male migrants, arriving to seek work, were required by city regulations to report to the Dispensaire for an exam. Anyone experiencing symptoms could visit the place voluntarily, and there was no charge for treatment. But the bulk of the caseload, according to Pepin, "consisted of thousands of asymptomatic free women who came for screening because they were required to do so by law, in theory every month." The colonial government accepted prostitution as an ineradicable fact but evidently hoped to keep the trade hygienic—so femmes libres were obliged to get checked.

If a person tested positive for syphilis or gonorrhea, he or she would be treated. But the diagnostic testing was imprecise. Any free woman or male migrant who had once been exposed to yaws (caused by a bacterium very similar to the syphilis bacterium, but not sexually transmissible) might flunk the blood test, be classed as syphilitic, and receive a long course of drugs containing arsenic or bismuth. Harmless vaginal flora could be mistaken for gonococcus, the agent of gonorrhea. A woman diagnosed gonorrheic might be injected with typhoid vaccine, or a drug called Gono-yatren, or (even Jacques Pepin seems puzzled by this one) milk. During the 1930s and 1940s, the Dispensaire Antivénérien administered more than 47,000 injections annually. Most were intravenous. Straight into the blood. With increased migration to the city following World War II, the numbers rose. In the early 1950s, the quackier remedies (intravenous milk?) and the metallic poisons gave way to penicillin and streptomycin, which had longer-lasting effects and therefore meant fewer shots. The campaign peaked in 1953, at about 146,800 injections, or roughly 400 per day. Many if not most of those injections were administered to femme libres, sex workers, ladies of hospitality, however you want to describe them, who had multiple male clients. They came and went. The syringes were rinsed and reused. This in a city where HIV had arrived.

Six years later came the blood sample that yielded the HIV sequence now known as ZR59. One year after that, DRC60. The virus had spread and diversified. It was at large. No one can say whether either of those two patients had ever visited the Dispensaire Antivénérien for a shot. But if they hadn't, they probably knew someone who had.

24

From this point the story gets huge and various, literally going off in all directions. It explodes out of Léopoldville like an infectious starburst. I won't try to trace those diverging trajectories—a task for ten other books, with purposes different from mine—but I'll sketch the pattern, then focus briefly on one that's especially notorious.

During its decades of inconspicuous transmission in Léopoldville, the virus continued to mutate (and probably also to recombine, mixing larger sections of genome from one virion to another), and those copying errors drove its diversification. Most mutations are insignificant changes, or else fatal mistakes, bringing the mutant to a dead end, but with so many billions of virions replicating, chance did provide a small, rich supply of viable new variants. The campaigns of injectable drug treatments, at the

Dispensaire Antivénérien and elsewhere, may have helped foster this process by transmitting the virus quickly into more human hosts and increasing its total population. The more virions, the more mutations; the more mutations, the more diversity.

The HIV-1 group M lineage became split into nine major subdivisions, which are now known as subtypes and labeled with letters: A, B, C, D, F, G, H, J, K. (Don't confuse those, if you can help it, with the eight groups of HIV-2, designated A through H. And why are E and I missing? Never mind why. Such edifices of labeling get built piecemeal, like slums of cardboard and tin, not with architectural forethought.) As time passed, as the human population of Léopoldville grew, as travel increased, viruses of those nine subtypes emerged from the city, radiating outward across Africa and the world. Some of them went by airplane and others by more mundane means of transport: bus, boat, train, bicycle, hitchhiking on a transcontinental truck. Foot. Subtype A got to East Africa, probably via the city of Kisangani, halfway between Léopoldville and Nairobi. Subtype C spread to southern Africa, probably via Lubumbashi, way down in the Congolese southeast. Seeping across Zambia, achieving rapid transmission in mining towns full of workers and prostitutes, subtype C proliferated throughout South Africa, Mozambique, Lesotho, and Swaziland. It went on to India, which is linked to South Africa by channels of exchange as old as the British empire, and to East Africa. Subtype D established itself alongside subtypes A and C in the countries of East Africa, except for Ethiopia, which for some reason became afflicted early and almost exclusively with subtype C. Subtype G got up into West Africa. Subtypes H, J, and K remained mostly in Central Africa, from Angola to the Central African Republic. In all these places, after the usual lag of years between infection and full-blown AIDS, people began dying. And then there's subtype B.

Sometime around 1966, subtype B crossed from Léopoldville to Haiti.

How it did that is unknown, and can probably never be known, but Jacques Pepin's archival burrowing provides new support for one plausible old scenario. When the Belgian government abruptly relinquished its African colony, on June 30, 1960, under the stern encouragement of Patrice Lumumba and his Mouvement National Congolais, tens of thousands of Belgian expatriates—almost an entire middle class of civil servants, teachers, doctors, nurses, technical experts, and business managers—found themselves unwelcome and uncomfortable in the new republic, and they began flooding homeward. Crowding the planes for Brussels. Their departure created a vacuum, since the Belgian regime had pointedly avoided educating its colonial subjects. There wasn't a single Congolese medical doctor, for instance. Few teachers. The country suddenly needed help. The World Health Organization responded, sending physicians, and the United Nations (through its Educational, Scientific, and Cultural Organization, UNESCO) also began enlisting skilled people to work in Congo: teachers, lawyers, agronomists, postal administrators, and other bureaucrats, technicians, and professionals. Many of those recruits came from Haiti. It was a natural fit: The Haitians spoke French as did the Congolese, they came from African roots, they had education but very little opportunity at home under the dictatorship of Papa Doc Duvalier.

During the first year of independence, half the teachers sent by UNESCO to Congo were Haitians. By 1963, according to one estimate, a thousand Haitians were employed in the country. Another estimate says that a total of forty-five hundred Haitians served hitches in Congo during the 1960s. Evidently there's no surviving, authoritative manifest. Anyway, lots of Haitians, multiple thousands. Some brought families, some came alone.

Among the single men, we can assume, few remained celibate. Most of them probably had Congolese girlfriends or visited femmes libres. For a few years it may have been a good life. But the Haitians were less needed and less welcome as Congo began training its own people, especially after Joseph Désiré Mobutu seized power in 1965. Less still when, in the early 1970s, he changed his name to Mobutu Sese Seko (roughly, "the all-powerful warrior"), changed his country's name to Zaire, and announced a policy of *Zaireanisation*. Many or most of the Haitians, during those years, went home. Their time of being useful and appreciated black brothers from the Americas had passed.

At least one of those returnees, probably among the earliest of them, seems to have carried HIV.

More specifically: Someone brought back to Haiti, along with Congolese memories, a dose of HIV-1 group M subtype B.

You can see where this is going, but you might not expect how it gets there. Jacques Pepin's research has shed some new light on what may have happened in Haiti during the late 1960s and early 1970s to multiply and forward the virus. One thing that happened was that, from a single HIV-positive person in 1966 or thereabouts, the virus spread fast through the Haitian population. Evidence for that spread came later, from blood samples given by 533 young mothers in a Port-au-Prince slum, who agreed in 1982 to participate in a measles study at a local pediatric clinic. Tested retrospectively, those samples revealed that 7.8 percent of the women had been HIV-positive. That number was startlingly high, for such a newly arrived virus, and caused Pepin to suspect that "there must have been a very effective amplification mechanism" operating in Haiti during the early years—more effective than sex. He found a candidate: the blood plasma trade.

Plasma, the liquid component of blood (minus the cells), is valuable stuff for its antibodies and albumin and clotting factors. Demand for it rose sharply during the period around 1970, and to meet the demand a process called plasmapheresis was developed. Plasmapheresis entails drawing blood from a donor, separating the cells from the plasma by means of filtering or centrifuging, putting the cells back into the donor, and keeping the plasma as a harvested product. One advantage of this process is that it allows donors (who are usually in fact sellers, paid for their trouble and needing the money) to be tapped often rather than just a couple times per year. Giving up your plasma, for the good of others or for profit, doesn't leave you anemic. You can go back and give again the following week. One disadvantage of the procedure—and it's a huge one, but wasn't recognized in the early days—is that a plasmapheresis machine, gargling your blood and the blood of many other donors over the course of days, can infect you with a blood-borne virus.

This happened to hundreds of paid plasma donors in Mexico during the mid-1980s. It happened to a quarter million luckless donors in China. Jacques Pepin thinks it happened in Haiti too.

He found reports of a plasmapheresis center in Port-au-Prince, a private business known as Hemo Caribbean, that operated profitably during 1971 and 1972. It was owned by an American investor, a man named Joseph B. Gorinstein, based in Miami, with links to the Haitian Minister of the Interior. Donors received three dollars per liter. Their vitals were checked before they could sell plasma, but of course nobody screened them for HIV—which didn't yet exist as an acronym, or an infamous global scourge, only as a quiet little virus that lived in blood. According to an article that ran in *The New York Times* on January 28, 1972, Hemo Caribbean was then exporting between five and six thousand liters of frozen blood plasma to the United States each month. The wholesale customers were American companies, which marketed the product for use in transfusions, tetanus shots, and other medical applications. Mr. Gorinstein wasn't available for comment.

Papa Doc had meanwhile died, in 1971, and been succeeded by his son Jean-Claude (Baby Doc) Duvalier. Annoyed by the *Times* publicity, Baby Doc ordered that Gorinstein's plasmapheresis center be closed. The Haitian Catholic Church condemned the blood trade as exploitation. Beyond that, the story of Hemo Caribbean drew little notice at the time. No one yet realized how devastating blood-product contamination could be. Nor did the CDC's *Morbidity and Mortality Weekly Report* mention it, a decade later, when breaking the news that Haitians seemed especially at risk for the mysterious new immune-deficiency syndrome. Randy Shilts didn't mention it in *And the Band Played On*. The only allusion to Haitian blood plasma that I recall, from the years before Jacques Pepin's book, came during my conversation with Michael Worobey in Tucson.

Shortly before publishing on DRC60 and ZR59, Worobey coauthored another notable paper, dating the emergence of HIV in the Americas. The first author was a postdoc named Tom Gilbert, in Worobey's lab, and in the anchor position was Worobey himself. This was the work, based on analyses of viral fragments from archived blood cells, that placed the arrival of HIV in Haiti to about 1966. It appeared in the *Proceedings of the National Academy of Sciences*. Soon afterward, Worobey got a peculiar e-mail from a stranger. Not a scientist, just someone who had caught wind of the subject. A reader of newspaper coverage, a listener to radio. "I think he was from Miami," Worobey told me. "He said he worked in an airport that dealt with the blood trade." The man had certain memories. Maybe they haunted him. He wanted to share them. He wanted to tell Worobey about cargo planes arriving full of blood.

25

The next leap of the virus was small in distance and large in consequence. Port-au-Prince is just seven hundred miles from Miami. A ninety-minute flight. Part of the project that Tom Gilbert undertook, in Worobey's lab, was to date when HIV had arrived in the United States. To do that he needed samples of old blood. Whether the blood had reached America in bottles, in bags, or in immigrant Haitians didn't much matter for this purpose.

Worobey, serving as Gilbert's adviser, remembered a study of immunodeficient Haitian immigrants, published twenty years earlier. (I alluded to the same study at the beginning of this book, quoting its observation of something that seemed "strikingly similar" to the new syndrome of immunodeficiency among American homosexuals.) That study had been led by a physician named Arthur E. Pitchenik, working at Jackson Memorial Hospital in Miami. Pitchenik was an expert on tuberculosis, and beginning in 1980 he noticed an unusual incidence of that disease, as well as *Pneumocystis* pneumonia, among Haitian patients. He had sounded the first alarm about Haitians as a risk group for the new immune-deficiency syndrome, alerting the CDC. In the course of clinical work and research, Pitchenik and his colleagues drew blood from patients and centrifuged it, separating serum from cells, so they could look at certain types of lymphocyte. They also froze some samples, on the assumption those might be useful to other researchers later. They were right. But for a long time no one seemed interested. Then, after two decades, Arthur Pitchenik got a call from Michael Worobey in Tucson. Yes, Pitchenik said, he would be glad to send some material.

Worobey's lab received six tubes of frozen blood cells, and Tom Gilbert managed to amplify viral fragments from five. Those fragments, after genetic sequencing, could be placed into context as limbs on another family tree—just as Worobey himself would later do with DRC60 and ZR59, and as Beatrice Hahn's group was doing with SIV_{cpz}. It was molecular phylogenetics at work. In this case, the tree represented the diversified lineage of HIV-1 group M subtype B. Its major limbs represented the virus as known from Haiti. One of those limbs encompassed a branch from which grew too many small twigs to portray. So in the figure as eventually published, that branch and its twigs were blurred—depicted simply as a solid cone of brown, like a sepia shadow, within which appeared a list of names. The names told where subtype B had gone, after passing through Haiti: the United States, Canada, Argentina, Colombia, Brazil, Ecuador, the Netherlands, France, the United Kingdom, Germany, Estonia, South Korea, Japan, Thailand, and Australia. It had also bounced back to Africa. It was HIV globalized.

This study by Gilbert and Worobey and their colleagues delivered one other piquant finding. Their data and analysis indicated that just a single migration of the virus—one infected person or one container of plasma—accounted for bringing AIDS to America. That sorry advent had occurred in 1969, plus or minus about three years.

So it lurked here for more than a decade before anyone noticed. For more than a decade, it infiltrated networks of contact and exposure. In particular, it followed certain paths of chance and opportunity into certain subcategories of the American populace. It was no longer a chimpanzee virus. It had found a new host and adapted, succeeding brilliantly, passing far beyond the horizons of its old existence within *Pan troglodytes*. It reached hemophiliacs through the blood supply. It reached drug addicts through shared needles. It reached gay men—reached deeply and devastatingly into their circles of love and acquaintance—by sexual transmission, possibly from an initial contact between two males, an American and a Haitian.

For a dozen years it traveled quietly from person to person. Symptoms were slow to arise. Death lagged some distance behind. No one knew. This virus was patient, unlike such hasty, peremptory bugs as Ebola and Marburg and the one that causes SARS. More patient even than rabies, but equally lethal. Somebody gave it to Gaëtan Dugas. Somebody gave it to Randy Shilts. Somebody gave it to a thirty-three-year-old Los Angeles man, who eventually fell ill with pneumonia and a weird oral fungus and, in March 1981, walked into the office of Dr. Michael Gottlieb.

NOTES

A number of the principals who feature in this account gave me their generous cooperation by way of sitting for interviews or responding to questions by e-mail or phone: Robert Gallo, Jane Goodall, Beatrice Hahn, Jean-Marie Kabongo, Phyllis Kanki, Brandon Keele, Elizabeth Lonsdorf, Martin Muller, J. J. Muyembe, Martine Peeters, Jane Raphael, Dirk Teuwen, Karen Terio, and Michael Worobey. Direct sourcing from those conversations, and from my field reporting, is clear from context. These notes refer to printed sources.

18. *Gottlieb's barebones text*: Gottlieb et al. (1981), 250.
19. Morbidity and Mortality Weekly Report *carried Friedman-Kien's communication*: Friedman-Kien. (1981), 305–306.
19. *saw a "syndrome" that seemed "strikingly similar"*: Pitchenik et al. (1983), 353–354.
19. *who became notorious as "Patient Zero"*: Shilts (1987), 23. But see also Auerbach et al. (1984), 489.
19. *as the man who "carried the virus out of Africa"*: e.g., http://en.wikipedia.org/wiki/Gaëtan_Dugas.

20. *even "gorgeous" in some eyes*: Shilts (1987), 21, 47.
20. *Dugas himself reckoned*: Shilts (1987), 83.
20. *and say: "I've got gay cancer"*: Shilts (1987), 165.
20. *"Although the cause of AIDS is unknown"*: Auerbach et al. (1984), 490.
21. *Randy Shilts later transformed*: Shilts (1987), 23.
21. *HIV had already arrived in North America when*: Gilbert et al. (2007), 18566, 18568.
22. *A Danish doctor named Grethe Rask*: Shilts (1987), 4–7; Bygbjerg (1983), 925.
22. *"I'd better go home to die."*: Shilts (1987), 6. Shilts seems to have interviewed Bygbjerg (but not Rask herself), as well as drawing from Bygbjerg (1983).
22. *Nine years later, a sample of Rask's blood serum*: Hooper (1999), 95, 879.
23. *GRID was one, standing for*: Shilts (1987), 121; Engel (2006), 6.
23. *Some doctors preferred ACIDS*: Shilts (1987), 138.
23. *"Kaposi's sarcoma and opportunistic infections"*: *Morbidity and Mortality Weekly Report,* June 11, 1982, 294.
23. *By September 1982,* MMWR *had switched*: *Morbidity and Mortality Weekly Report,* September 24, 1982, 507.
24. *Montagnier's research focused mainly*: Montagnier (2000), 27–30, 38, 47.
25. *"AIDS could not be caused by a conventional bacterium"*: Montagnier (2000), 42.
25. *The only known human retrovirus as of early 1981*: Gallo (1991), 91–93, 99.
25. *A related retrovirus, feline leukemia virus*: Barré-Sinoussi (2003a), 844.
25. *Montagnier's group in Paris, screening cells*: Barré-Sinoussi et al. (1983), 868; Montagnier (2000), 57. Barré-Sinoussi and Montagnier didn't name it LAV in the original 1983 paper, but slightly later.
26. *Gallo's group came up with*: Gallo et al. (1983), 865–866; Gallo (1991), 92–93, 99, 117. Gallo uses Arabic numerals (e.g., HTLV-

1, HTLV-2) in his book; but in the scientific papers, he and others use Roman numerals.

26. *He called this newest bug HTLV-III*: Gallo et al. (1984), 500, 502; Popovic et al. (1984), 497.
26. *An editorial in the same issue of* Science: Marx (1983), 806.
26. *Then again, neither was Gallo's*: Gallo and Montagnier (1988), 44; Gallo (1991), 186; Crewdson (2002), 163–166.
26. *Montagnier had personally delivered*: Montagnier (2000), 60–62, 68–69.
27. *Meanwhile the third team of researchers:* Crewdson (2002), 143, 158.
27. *"more than 4000 individuals in the world"*: Levy et al. (1984), 840.
27. *"Our data cannot reflect a contamination"*: Levy et al. (1984), 842.
28. *A distinguished committee of retrovirologists*: Crewdson (2002), 179–180, 236.
29. *There she saw a weird problem:* Essex and Kanki (1988), 67; Letvin et al. (1983), 2718–2719.
29. *they did find a new retrovirus*: Daniel et al. (1985); Kanki et al. (1985b).
29. *for what soon would be renamed HIV*: i.e., they referred to the AIDS virus as HTLV-III and called their macaque virus STLV-III.
29. *This discovery, they wrote*: Daniel et al. (1985) and Kanki et al. (1985b), last paragraph of each.
29. *Only a single sentence at the end*: Kanki et al. (1985b), 1201.
30. *Kanki and Essex looked at other Asian macaques*: Essex and Kanki (1988), 67–68.
30. *"In 1985, the highest rates"*: Essex and Kanki (1988), 68.
30. *Kanki grew isolates of live virus*: Kanki et al. (1985b), 952–953.
31. *"must have evolved mechanisms"*: Essex and Kanki (1988), 68.
31. *The samples arrived with coded labeling*: Essex and Kanki (1988), 69; Kanki et al. (1986), 238.
31. *Despite one possible misstep*: Kanki et al. (1986), 238; regarding contamination and confusion, cf. Montagnier (2000), 80–81; Hooper (1999), 108; Kestler et al. (1988), 619, and Essex and Kanki's reply to Kestler, same issue, 621–622; Barin et al. (1985), 1387.
32. *It more closely resembled SIV strains*: Barin et al. (1985), 1387.

32. *Montagnier and his colleagues screened the blood*: Montagnier (2000), 79–81; Clavel et al. (1986), 343–344.
32. *This man showed symptoms of AIDS*: Clavel et al. (1986), 343–344; Montagnier (2000), 79–80.
32. *Eventually, when all parties embraced the label*: Clavel (1986), 346; Montagnier (2000), 81.
33. *Possibly it was already with us*: See Fukasawa et al. (1988), 460; Mulder (1988), 396.
33. *when a group of Japanese researchers*: Fukasawa et al. (1988), 457.
33. *The nucleotide sequence of its retrovirus*: Fukasawa et al. (1988), 457, 459; Mulder (1988), 396.
34. *A commentary in the journal* Nature: Mulder (1988), 396.
35. *had noticed a leprosy-like infection*: Gormus et al. (2004), 216.
35. *not known to be transmissible from people*: Wolf et al. (1985), 529.
35. *The animal in question, a sooty mangabey*: Gormus (2004), 216. The story unfolds from Gormus (in retrospect) to Wolf et al. (1985) to Murphey-Corb et al. (1986).
36. *revealed that the virus was quite prevalent among them*: Murphey-Corb et al. (1986), 437.
36. *Other investigators soon found it too*: Hirsch et al. (1989), 389, and its citation notes 9–11.
37. *Now there were three known variants*: Kanki (1986); Daniel et al. (1985). The SIV name came later, however, after they stopped using HTLV and STLV.
37. *"These results suggest that SIV_{sm} has infected macaques"*: Hirsch et al. (1989), 389.
38. *HIV-2 is confined mostly to West African countries*: this sentence and the next three, Reeves and Doms (2002), 1254–1255.
39. *Peeters along with several associates was tasked*: this paragraph, Peeters et al. (1989), 625–626.
40. *announcing the new virus and calling it SIV_{cpz}*: Peeters et al. (1989), 625, 627. More precisely, they called it $SIV_{cpz\text{-}GAB\text{-}1,}$ indicating not just the new strain of virus but the identity of the specific isolate.
40. *"It has been suggested that human AIDS"*: Peeters et al. (1989), 629.

40. *In 1992 Peeters published another*: Peeters et al. (1992), 448.
41. *not a single one had yielded traces of* SIV_{cpz}: Sharp and Hahn (2010), 2488.
42. *by the year 2000 seven groups of HIV-2*: Reeves and Doms (2002), 1253.
43. *So did the later addition, group H*: Santiago et al. (2005), 12515.
43. *The eventual fourth kind, group P:* Plantier et al. (2009), 1–2.
43. *Scientists think that each of those twelve groups*: Reeves and Doms (2002), 1253 regarding HIV-2; Sharp and Hahn (2010), 2487, regarding HIV-1.
44. *In September of that year, a young print-shop worker*: Zhu and Ho (1995), 503; Hooper (1999), 21–22, 122 ff.
44. *Thirty-one years later, in the era of AIDS*: Corbitt et al. (1990), cited in Zhu and Ho (1995).
45. *must have reflected a laboratory mistake*: Zhu and Ho (1995), 503–504.
45. *A team of researchers including Tuofu Zhu*: Zhu and Ho (1995), 503–504.
45. *a small tube of blood plasma, drawn from a Bantu man*: Nahmias et al. (1986), 1279.
45. *the only one that tested unambiguously positive*: Zhu et al. (1998), 594.
46. *In their paper, published in February 1998*: Zhu et al. (1998).
46. *DRC60 was a biopsy specimen*: Worobey et al. (2008), 661.
50. *with a spillover as early as 1908*: Worobey et al. (2008), 661.
51. *This one was heterodox and highly controversial*: Hooper (2001), 803. Hooper presents somewhat different numbers in Hooper (1999), 265–277, 378–379.
51. *viral or bacterial contamination of a vaccine*: e.g., with SV40 in some of the Salk vaccines, Shah and Nathanson (1976), 3.
51. *a group of Italian children*: Willrich (2001), 181.
51. *Smallpox vaccine administered to kids in Camden*: Willrich (2001), 171–176, 192, 201.
51. *a batch of diphtheria antitoxin prepared in St. Louis*: Willrich (2001), 178.
51. *Formaldehyde was sometimes added*: Oshinsky (2006), 281.

52. *some of the early batches of the Salk polio vaccine*: Shah and Nathanson (1976), 2; Shah (2004), 2061.
52. *That the vaccine in question had been given to Africans*: Koprowski (2001); Plotkin (2001).
52. *Koprowski himself visited Stanleyville*: Hooper (1999), 267–273, 523–524.
52. *Children and adults lined up trustingly*: Hooper (1999), 268–269, 273–274.
52. *roughly seventy-five thousand kids*: Hooper (1999), 275.
53. *chimpanzee kidneys drawn from animals infected*: Hooper (2001), 803–805, versus Plotkin (2001), 815–816.
53. *The result of that flawed vaccinating*: Hooper (2001), 803.
53. *certain people have argued*: e.g., Hooper, Louis Pascal, William Hamilton, Tom Curtis.
53. *had put Tom Curtis onto the story*: i.e., Blaine Elswood, as mentioned in Curtis (1992), 3 (pagination of digital version).
53. *"The origin of the AIDS virus is of no importance"*: Curtis (1992), 21.
53. *"It's distracting, it's nonproductive"*: Curtis (1992), 21.
53. *lawyers for Hilary Koprowski filed a lawsuit*: Hooper (1999), 254, 456.
54. *"The controversy surrounding the source of the Nile"*: Hooper (1999), 4.
59. *he screened just 27 of the 813 tissue blocks*: Worobey (2008), 661, 663.
60. *They both fell within the range*: Worobey (2008), 661–662.
60. *differed by 12 percent between the two versions*: actually, 11.7 percent: Worobey (2008), 662.
60. *he placed the most recent common ancestor*: Worobey (2008), 663, Table 1.
60. *"Our estimation of divergence times"*: Worobey (2008), 663.
62. *"the most persuasive evidence yet"*: Weiss and Wrangham (1999), 385.
62. *their analysis of viral strains linked it*: Gao et al. (1999), 436–437.
62. *on viruses drawn from captive chimps*: the new chimp in Gao's data was Marilyn, captive in the United States: Gao et al. (1999), 437.

63. *Mario L. Santiago topped a list of coauthors*: Santiago et al. (2002), 465.
63. *he invented methods*: Santiago et al. (2002), 465.
65. *they collected 446 samples of chimpanzee dung*: Keele et al. (2006), 523.
66. *prevalence rates up to 35 percent*: Keele et al. (2006), 525, map on 523.
66. *a twig amid the same little branch*: Keele et al. (2006), 524–525, Figures 3 and 4.
67. *shockingly similar to HIV-1 group M*: Keele et al. (2006), 525.
68. *"We show here that the SIV_{cpzPtt} strain"*: Keele et al. (2006), 526.
68. *"In humans, direct exposure to animal blood"*: Hahn et al. (2000), 611.
68. *"The likeliest route of chimpanzee-to-human transmission"*: Sharp and Hahn (2010), 2492.
70. *Léopoldville contained fewer than ten thousand people*: Worobey (2008), 663, Figure 3 and its caption, citing Chitnis et al. (2000).
70. *"a hard mission field," according to one Swedish missionary*: Martin (2002), 20, 25.
70. *due to colonial policies that discouraged married men*: Pepin (2011), 70–73. The rest of this paragraph, and the next: Pepin (2011).
71. *a lively market in smoked fish*: Harms (1981), 229.
71. *Ivory, rubber, and slaves were traded there*: Harms (1981), 227–229.
72. *By 1940, its population had edged up*: Chitnis et al. (2000), 6.
74. *apes, elephants, lions, and a few other species were protected*: *Wildlife Justice*, No. 2 (May 2006), 8.
74. *Drori gave me a LAGA newsletter*: *Wildlife Justice*, No. 4 (November 2006).
74. *Drori's newsletter mentioned a raid*: *Wildlife Justice*, No. 4 (November 2006), 5.
75. *Another bust, against a dealer*: *Wildlife Justice*, No. 4 (November 2006), 5; *Wildlife Justice*, No. 2 (May 2006), 2, 12.
76. *a driver unloading chimpanzee arms and legs*: Peterson (2003), 46, 159.
76. *roughly 5 million metric tons of bushmeat*: Peterson (2003), 65.
80. *where Karl Amman saw chimpanzee arms stashed*: Peterson (2003), 46.

80. *Chimp fecal samples from hereabouts*: Keele et al. (2006), 525.
84. *possibly of the Mpiemu or the Kako*: Giles-Vernick (2002), 22.
94. *A study of bushmeat traffic in and around Ouesso*: this sentence and the rest of the paragraph: Hennessey and Rogers (2008), 179–183.
109. *the prevailing impression was that it's harmless in chimpanzees*: e.g., Novembre et al. (1997), 11748, 11752.
109. *When a single lab chimpanzee did progress to AIDS*: Novembre et al. (1997), 4086.
110. *"survived their own AIDS-like pandemic"*: Cohen (2002), 15. Cohen was reporting on de Groot et al. (2002).
110. *a naturally occurring infection in more than forty different species*: Sharp and Hahn (2010), 2487.
110. *it hasn't shown up among wild monkeys in either Asia*: Sharp and Hahn (2010), 2487.
110. *none of those SIVs seems to cause immunodeficiency*: Sharp and Hahn (2010), 2490.
110. *a close similarity between their respective SIVs*: Sharp et al. (2005), 3893.
111. *That length of time would allow divergence*: Sharp et al. (2005), 3893.
111. *noticed that SIV_{cpz} seems to be a hybrid virus*: Bailes et al. (2003), 1713.
111. *Possibly just hundreds of years ago*: Wertheim and Worobey (2009), 5–6; Pepin (2011), 41, citing Wertheim and Worobey (2009).
112. *What the Nottingham group suggested*: Bailes (2003), 1713.
124. *these all came together in a paper*: Keele et al. (2009), 515.
126. *a series of well-intended campaigns*: Pepin (2011), 117; Pepin and Labbé (2008).
127. *2 million syringes were produced globally in 1930*: Drucker et al. (2001), 1989. See also Marx et al. (2001), 914.
127. *treated 5,347 trypanosomiasis cases*: Pepin (2011), 122, 163.
127. *"The Congo contains various health institutions"*: Beheyt (1953), quoted in Pepin (2011), 164.
128. *"The large number of patients"*: Beheyt (1953), quoted in Pepin (2011), 164.

128. *performed 207,089 injections of tryparsamide*: this sentence and the rest of the paragraph: Pepin (2011), 125–128.
128. *doubt that needles were necessary in any such way*: Worobey (2008), in Volberding et al. (2008), 18.
129. *It dates back more than a decade*: Marx et al. (2001), 911.
129. *Jacques Pepin picked up where Preston Marx left off*: Pepin and Frost (2011), 421–422.
130. *a clinic known as the Dispensaire Antivénérien*: Pepin (2011), 160.
130. *"consisted of thousands of asymptomatic free women"*: Pepin (2011), 161.
130. *Any free woman or male migrant*: this sentence and the rest of the paragraph, Pepin (2011), 160–163.
132. *The HIV-1 group M lineage became split*: Taylor et al. (2008), 1591; Worobey (2008), in Volberding (2008), 15.
132. *Subtype A got to East Africa*: Pepin (2011), 212–213.
132. *Subtype D established itself alongside subtypes A and C*: Taylor et al. (2008), 1595, Table 2; Hemelaar et al. (2006), W17, Table 2, and W18, Table 3.
133. *subtype B crossed from Léopoldville to Haiti*: Gilbert et al. (2007), 18566, 18568, Figure 2.
133. *new support for one plausible old scenario*: The recognition of Haitian professionals having gone to Congo after Independence dates back at least to Shilts (1987), 392–393. This paragraph and the next: Pepin (2011), 187–190.
134. *Someone brought back to Haiti*: Gilbert et al. (2007), 18566.
134. *those samples revealed that 7.8 percent of the women*: Boulos et al. (1990), 7222–7223, and cited in Pepin (2011), 196.
134. *"there must have been a very effective amplification mechanism"*: Pepin (2011), 196.
135. *hundreds of paid plasma donors in Mexico*: Pepin (2011), 199.
135. *a quarter million luckless donors in China*: Pepin (2011), 200.
135. *reports of a plasmapheresis center in Port-au-Prince*: Pepin (2011), 201–202; Severo (1972). Pepin cites Severo, but I used Severo directly. Pepin hyphenates the name, Hemo-Caribbean, but Severo doesn't and there's still such a company, listed online as Hemo Caribbean.

136. *ordered that Gorinstein's plasmapheresis center be closed*: Pepin (2011), 202.
136. *Nor did the CDC's* Morbidity and Mortality Weekly Report *mention it*: *Morbidity and Mortality Weekly Report*, July 9, 1982, 31(26): 354.
136. *Randy Shilts didn't mention it*: though he came close, discussing Haitians and blood, e.g., Shilts (1987), 135.
137. *beginning in 1980 he noticed*: Pitchenik et al. (1983), 277, 278, table.
137. *He had sounded the first alarm about Haitians*: *Morbidity and Mortality Weekly Report*, July 9, 1982, 31(26): 354ff.
138. *Tom Gilbert managed to amplify*: Gilbert et al. (2007), 18569.
139. *walked into the office of Dr. Michael Gottlieb*: Gottlieb et al. (1981), 250.

BIBLIOGRAPHY

Auerbach, D. M., W. W. Darrow, H. W. Jaffe, and J. W. Curran. 1984. "Cluster of Cases of the Acquired Immune Deficiency Syndrome. Patients Linked by Sexual Contact." *The American Journal of Medicine,* 76 (3).

Bailes, E., F. Gao, F. Biboilet-Ruche, V. Courgnaud, M. Peeters, P. A. Marx, B. H. Hahn, and P. M. Sharp. 2003. "Hybrid Origin of SIV in Chimpanzees." *Science,* 300.

Barin, F., S. M'Boup, F. Denis, P. Kanki, J. S. Allan, T. H. Lee, and M. Essex. 1985. "Serological Evidence for Virus Related to Simian T-Lymphotropic Retrovirus III in Residents of West Africa." *The Lancet,* 2.

Barré-Sinoussi, F. 2003a. "The Early Years of HIV Research: Integrating Clinical and Basic Research." *Nature Medicine,* 9 (7).

———. 2003b. "Barré-Sinoussi Replies." *Nature Medicine,* 9 (7).

Barré-Sinoussi, F., J. C. Cherrmann, F. Rey, M. T. Nugeyre, S. Chamaret, J. Gruest, C. Dauguet, et al. 1983. "Isolation of a T-Lymphotropic Retrovirus from a Patient at Risk for Acquired Immune Deficiency Syndrome (AIDS)." *Science,* 220.

Beheyt, P. 1953. "*Contribution à l'étude des hepatites en Afrique. L'hépatite épidémique et l'hépatite par inoculation.*" *Annales de la Société Belge de Médicine Tropicale, 33.*

Boulos, R., N. A. Halsey, E. Holt, A. Ruff, J. R. Brutus, T. C. Quin, M. Adrien, and C. Boulos. 1990. "HIV-1 in Haitian Women 1982–1988." *Journal of Acquired Immune Deficiency Syndromes,* 3.

Bygbjerg, I. C. 1983. "AIDS in a Danish Surgeon (Zaire, 1976)." *The Lancet,* 1 (2).

Chitnis, A., D. Rawls, and J. Moore, et al. 2000. "Origin of HIV Type 1 in Colonial French Equatorial Africa?" *AIDS Research and Human Retroviruses,* 16 (1).

Clavel, F., D. Guétard, F. Brun-Vézinet, S. Chamaret, M. A. Rey, M. O. Santos-Ferreira, A. G. Laurent, et al. 1986. "Isolation of a New Human Retrovirus from West African Patients with AIDS." *Science,* 233.

Cohen, P. 2002. "Chimps Have Already Conquered AIDS." *New Scientist,* August 24.

Crewdson, J. 2002. *Science Fictions: A Scientific Mystery, a Massive Coverup, and the Dark Legacy of Robert Gallo.* Boston: Little, Brown.

Curtis, T. 1992. "The Origin of AIDS." *Rolling Stone,* March 19.

Daniel, M. D., N. L. Letvin, N. W. King, M. Kannagi, P. K. Sehgal, R. D. Hunt, P. J. Kanki, et al. 1985. "Isolation of T-Cell Tropic HTLV-III-like Retrovirus from Macaques." *Science,* 228.

De Groot, N. G., N. Otting, G. G. Doxiadis, S. S. Balla-Jhagjoorsingh, J. L. Heeney, J. J. van Rood, P. Gagneux, et al. 2002. "Evidence for an Ancient Selective Sweep in the MHC Class I Gene Repertoire of Chimpanzees." *Proceedings of the National Academy of Sciences,* 99 (18).

Drucker, E., P. C. Alcabes, and P. A. Marx. 2001. "The Injection Century: Massive Unsterile Injections and the Emergence of Human Pathogens." *The Lancet,* 358.

Duesberg, P. 1996. *Inventing the AIDS Virus.* Washington, D.C.: Regnery Publishing.

Engel, J. 2006. *The Epidemic: A Global History of AIDS.* New York: Smithsonian Books/HarperCollins.

Epstein, H. 2007. *The Invisible Cure: Why We Are Losing the Fight Against AIDS in Africa.* New York: Picador.

Essex, M., and P. J. Kanki. 1988. "The Origins of the AIDS Virus." *Scientific American,* 259 (4).

Essex, M., S. M'Boup, P. J. Kanki, R. G. Marlink, and S. D. Tlou, eds. 2002. *AIDS in Africa.* 2nd ed. New York: Kluwer Academic/Plenum Publishers.

Friedman-Kien, A. E. 1981. "Disseminated Kaposi's Sarcoma Syndrome in Young Homosexual Men." *Journal of the American Academy of Dermatology,* 5.

Fukasawa, M., T. Miura, A. Hasegawa, S. Morikawa, H. Tsujimoto, K. Miki, T. Kitamura, and M. Hayami. 1988. "Sequence of Simian Immunodeficiency Virus from African Green Monkey, A New Member of the HIV/SIV Group." *Nature,* 333.

Gallo, R. 1991. *Virus Hunting: AIDS, Cancer, and the Human Retrovirus: A Story of Scientific Discovery.* New York: Basic Books.

Gallo, R. C., and L. Montagnier. 1988. "AIDS in 1988." *Scientific American,* 259 (4).

Gallo, R. C., S. Z. Salahuddin, M. Popovic, G. M. Shearer, M. Kaplan, B. F. Haynes, T. J. Palker, et al. 1984. "Frequent Detection and Isolation of Cytopathic Retroviruses (HTLV-III) from Patients with AIDS and at Risk for AIDS." *Science,* 224.

Gallo, R. C., P. S. Sarin, E. P. Gelmann, M. Robert-Guroff, E. Richardson, V. S. Kalyanaraman, D. Mann, et al. 1983. "Isolation of Human T-Cell Leukemia Virus in Acquired Immune Deficiency Syndrome (AIDS)." *Science,* 220.

Gao, F., E. Bailes, D. L. Robertson, Y. Chen, C. M. Rodenburg, S. F. Michael, L. B. Cummins, et al. 1999. "Origin of HIV-1 in the Chimpanzee *Pan troglodytes troglodytes.*" *Nature,* 397.

Gilbert, M. T. P., A. Rambaud, G. Wlasiuk, T. J. Spira, A. E. Pitchenik, and M. Worobey. 2007. "The Emergence of HIV/AIDS in the Americas and Beyond." *Proceedings of the National Academy of Sciences,* 104 (47).

Giles-Vernick, T. 2002. *Cutting the Vines of the Past: Environmental Histories of the Central African Rain Forest.* Charlottesville: University Press of Virginia.

Gormus, B. J., L. N. Martin, and G. B. Baskin. 2004. "A Brief History of the Discovery of Natural Simian Immunodeficiency Virus (SIV) Infections in Captive Sooty Mangabey Monkeys." *Frontiers in Bioscience,* 9.

Gottlieb, M. S., H. M. Shankar, P. T. Fan, A. Saxon, J. D. Weisman, and I. Pozalski. 1981. "*Pneumocystis* Pneumonia—Los Angeles." *Morbidity and Mortality Weekly Report*, June 5.

Hahn, B. H., G. M. Shaw, K. M. De Cock, and P. M. Sharp. 2000. "AIDS as a Zoonosis: Scientific and Public Health Implications." *Science*, 287.

Harms, R. W. 1981. *River of Wealth, River of Sorrow: The Central Zaire Basin in the Era of the Slave and Ivory Trade, 1500–1891.* New Haven: Yale University Press.

Hemelaar, J., E. Gouws, P. D. Ghys, and S. Osmanov. 2006. "Global and Regional Distribution of HIV-1 Genetic Subtypes and Recombinants in 2004." *AIDS*, 20 (16).

Hennessey, A. B., and J. Rogers. 2008. "A Study of the Bushmeat Trade in Ouesso, Republic of Congo." *Conservation and Society*, 6 (2).

Hirsch, V. M., R. A. Olmsted, M. Murphy-Corb, R. H. Purcell, and P. R. Johnson. 1989. "An African Primate Lentivirus (SIV_{sm}) Closely Related to HIV-2." *Nature*, 339.

Holmes, E. C. 2009. *The Evolution and Emergence of RNA Viruses.* Oxford: Oxford University Press.

Hooper, E. 1990. *Slim: A Reporter's Own Story of AIDS in East Africa*. London: The Bodley Head.

———. 1999. *The River: A Journey to the Source of HIV and AIDS.* Boston: Little, Brown.

———. 2001. "Experimental Oral Polio Vaccines and Acquired Immune Deficiency Syndrome." *Philosophical Transactions of the Royal Society of London*, 356.

Kanki, P. J., J. Alroy, and M. Essex. 1985a. "Isolation of T-Lymphotropic Retrovirus Related to HTLV-III/LAV from Wild-Caught African Green Monkeys." *Science*, 230.

Kanki, P. J., F. Barin, S. M'Boup, J. S. Allan, J. L. Romet-Lemonne, R. Marlink, M. F. Maclane, et al. 1986. "New Human T-Lymphotropic Retrovirus Related to Simian T-Lymphotropic Virus Type III ($STVL\text{-}III_{AGM}$)." *Science*, 232.

Kanki, P. J., M. F. Maclane, N. W. King Jr., N. L. Letvin, R. D. Hunt, P. Sehgal, M. D. Daniel, et al. 1985b. "Serologic Identification and Characterization of a Macaque T-Lymphotropic Retrovirus Closely Related to HTLV-III." *Science*, 228.

Keele, B. F., F. Van Heuverswyn, Y. Li, E. Bailes, J. Takehisa, M. L. Santiago, F. Bibollet-Ruche, et al. 2006. "Chimpanzee Reservoirs of Pandemic and Nonpandemic HIV-1." *Science,* 313.

Keele, B. F., J. Holland Jones, K. A. Terio, J. D. Estes, R. S. Rudicell, M. L. Wilson, Y. Li, et al. 2009. "Increased Mortality and AIDS-like Immunopathology in Wild Chimpanzees Infected with SIVcpz." *Nature,* 460.

Kermack, W. O., and A. G. McKendrick. 1927. "A Contribution to the Mathematical Theory of Epidemics." *Proceedings of the Royal Society of London,* A, 115.

Kestler, H. W. III, Y. Li, Y. M. Naidu, C. V. Butler, M. F. Ochs, G. Jaenel, N. W. King, et al. 1988. "Comparison of Simian Immunodeficiency Virus Isolates." *Nature,* 331.

Koprowski, H. 2001. "Hypothesis and Facts." *Philosophical Transactions of the Royal Society of London,* 356.

Korber, B., M. Muldoon, J. Theiler, F. Gao, R. Gupta, A. Lapedes, B. H. Hahn, et al. 2000. "Timing the Ancestor of the HIV-1 Pandemic Strains." *Science,* 288.

Letvin, N. L., K. A. Eaton, W. R. Aldrich, P. K. Sehgal, B. J. Blake, S. F. Schlossman, N. W. King, and R. D. Hunt. 1983. "Acquired Immunodeficiency Syndrome in a Colony of Macaque Monkeys." *Proceedings of the National Academy of Sciences,* 80.

Levy, J. A., A. D. Hoffman, S. M. Kramer, J. A. Landis, J. M. Shimabukuro, and L. S. Oshiro. 1984. "Isolation of Lymphocytopathic Retroviruses from San Francisco Patients with AIDS." *Science,* 225.

McNeill, W. H. 1976. *Plagues and Peoples.* New York: Anchor Books.

Martin, P. M. 2002. *Leisure and Society in Colonial Brazzaville.* Cambridge: Cambridge University Press.

Marx, J. L. 1983. "Human T-Cell Leukemia Virus Linked to AIDS." *Science,* 220.

Marx, P. A., P. G. Alcabes, and E. Drucker. 2001. "Serial Human Passage of Simian Immunodeficiency Virus by Unsterile Injections and the Emergence of Epidemic Human Immunodeficiency Virus in Africa." *Philosophical Transactions of the Royal Society of London,* 356.

May, R. 2001. "Memorial to Bill Hamilton." *Philosophical Transactions of the Royal Society of London,* 356.

Montagnier, L. 2000. *Virus: The Co-Discoverer of HIV Tracks Its Rampage and Charts the Future.* Translated from the French by Stephen Sartelli. New York: W. W. Norton.

———. 2003. "Historical Accuracy of HIV Isolation." *Nature Medicine,* 9 (10).

Morse, S. S., ed. 1993. *Emerging Viruses.* New York: Oxford University Press.

Mulder, C. 1988. "Human AIDS Virus Not from Monkeys." *Nature,* 333.

Murphey-Corb, M., L. N. Martin, S. R. Rangan, G. B. Baskin, B. J. Gormus, R. H. Wolf, W. A. Andres, et al. 1986. "Isolation of an HTLV-III-related Retrovirus from Macaques with Simian AIDS and Its Possible Origin in Asymptomatic Mangabeys." *Nature,* 321.

Nahmias, A. J., J. Weiss, X. Yao, F. Lee, R. Kodsi, M. Schanfield, T. Matthews, et al. 1986. "Evidence for Human Infection with an HTLV III/LAV-like Virus in Central Africa, 1959." *The Lancet,* 1, (8492).

Novembre, F. J., M. Saucier, D. C. Anderson, S. A. Klumpp, S. P. O'Neil, C. R. Brown II, C. E. Hart, et al. 1997. "Development of AIDS in a Chimpanzee Infected with Human Immunodeficiency Virus Type 1." *Journal of Virology,* 71 (5).

Oldstone, M.B.A. 1998. *Viruses, Plagues, and History.* New York: Oxford University Press.

Oshinsky, D. M. 2006. *Polio: An American Story.* Oxford: Oxford University Press.

Peeters, M., C. Honoré, T. Huet, L. Bedjabaga, S. Ossari, P. Bussi, R. W. Cooper, and E. Delaporte. 1989. "Isolation and Partial Characterization of an HIV-related Virus Occurring Naturally in Chimpanzees in Gabon." *AIDS,* 3 (10).

Peeters, M., K. Fransen, E. Delaporte, M. Van den Haesevelde, G. M. Gershy-Damet, L. Kestens, G. van der Groen, and P. Piot. 1992. "Isolation and Characterization of a New Chimpanzee Lentivirus (Simian Immunodeficiency Virus Isolate cpz-ant) from a Wild-Captured Chimpanzee." *AIDS,* 6 (5).

Pepin, J. 2011. *The Origins of AIDS.* Cambridge: Cambridge University Press.

Pepin, J., and E. H. Frost. 2011. "Reply to Marx et al." *Clinical Infectious Diseases*, Correspondence 52.

Pepin, J., and A.-C. Labbé. 2008. "Noble Goals, Unforeseen Consequences: Control of Tropical Diseases in Colonial Central Africa and the Iatrogenic Transmission of Blood-borne Diseases." *Tropical Medicine and International Health,* 13 (6).

Pepin, J., A.-C. Labbé, F. Mamadou-Yaya, P. Mbélesso, S. Mbadingaï, S. Deslandes, M.-C. Locas, and E. Frost. 2010. "Iatrogenic Transmission of Human T Cell Lymphotropic Virus Type 1 and Hepatitis C Virus through Parenteral Treatment and Chemoprophylaxis of Sleeping Sickness in Colonial Equatorial Africa." *Clinical Infectious Diseases,* 51.

Pepin, K. M., S. Lass, J. R. Pulliam, A. F. Read, and J. O. Lloyd-Smith. 2010. "Identifying Genetic Markers of Adaptation for Surveillance of Viral Host Jumps." *Nature,* 8.

Peterson, D. 2003. *Eating Apes.* With an afterword and photographs by Karl Ammann. Berkeley: University of California Press.

Pisani, E. 2009. *The Wisdom of Whores: Bureaucrats, Brothels, and the Business of AIDS.* New York: W. W. Norton.

Pitchenik, A. E., M. A. Fischl, G. M. Dickinson, D. M. Becker, A. M. Fournier, M. T. O'Connell, R. D. Colton, and T. J. Spira. 1983. "Opportunistic Infections and Kaposi's Syndrome Among Haitians: Evidence of a New Acquired Immunodeficiency State." *Annals of Internal Medicine,* 98 (3).

Plantier, J. C., M. Leoz, J. E. Dickerson, F. De Oliveira, F. Cordonnier, V. Lemée, F. Damond, et al. 2009. "A New Human Immunodeficiency Virus Derived from Gorillas." *Nature Medicine,* 15.

Plotkin, S. A. 2001. "Untruths and Consequences: The False Hypothesis Linking CHAT Type 1 Polio Vaccination to the Origin of Human Immunodeficiency Virus." *Philosophical Transactions of the Royal Society of London,* 356.

Popovic, M., M. G. Sarngadharan, E. Read, and R. C. Gallo. 1984. "Detection, Isolation, and Continuous Production of Cytopathic Retroviruses (HTLV-III) from Patients with AIDS and Pre-AIDS." *Science,* 224.

Reeves, J. D., and R. W. Doms. 2002. "Human Immunodeficiency Virus Type 2." *Journal of General Virology,* 83.

Santiago, M. L., C. M. Rodenburg, S. Kamenya, F. Bibollet-Ruche, F. Gao, E. Bailes, S. Meleth, et al. 2002. "SIVcpz in Wild Chimpanzees." *Science,* 295.

Santiago, M. L., F. Range, B. F. Keele, Y. Li, E. Bailes, F. Bibollet-Ruche, C. Fruteau, et al. 2005. "Simian Immunodeficiency Virus Infection in Free-Ranging Sooty Mangabeys (*Cercocebus atys atys*) from the Taï Forest, Côte d'Ivoire: Implications for the Origin of Epidemic Human Immunodeficiency Virus Type 2." *Journal of Virology,* 79 (19).

Severo, R. 1972. "Impoverished Haitians Sell Plasma for Use in the U.S." *The New York Times,* January 28.

Shah, K. V. 2004. "Simian Virus 40 and Human Disease." *Journal of Infectious Diseases,* 190.

Shah, K., and N. Nathanson. 1976. "Human Exposure to SV40: Review and Comment." *American Journal of Epidemiology,* 103 (1).

Sharp, P. M., and B. H. Hahn. 2010. "The Evolution of HIV-1 and the Origin of AIDS." *Philosophical Transactions of the Royal Society,* 365.

Sharp, P., G. Shaw, and B. Hahn. 2005. "Simian Immunodeficiency Virus Infection of Chimpanzees." *Journal of Virology,* 79(7).

Shilts, R. 1987. *And the Band Played On: Politics, People, and the AIDS Epidemic.* New York: St Martin's Griffin.

Sompayrac, L. 2002. *How Pathogenic Viruses Work.* Sudbury, MA: Jones and Bartlett Publishers.

Taylor, B. S., M. E. Sobieszczyk, F. E. McCutchan, and S. M. Hammer. 2008. "The Challenge of HIV-1 Subtype Diversity." *New England Journal of Medicine,* 358 (15).

Volberding, P. A., M. A. Sande, J. Lange, W. C. Greene, and J. E. Gallant, eds. 2008. *Global HIV/AIDS Medicine.* Philadelphia: Saunders Elsevier.

Weiss, R. A. 1988. "A Virus in Search of a Disease." *Nature,* 333.

———. 2001. "The Leeuwenhoek Lecture 2001. Animal Origins of Human Infectious Disease." *Philosophical Transactions of the Royal Society of London,* B, 356.

Weiss, R. A., and J. L. Heeney. 2009. "An Ill Wind for Wild Chimps?" *Nature,* 460.

Weiss, R. A., and R. W. Wrangham. 1999. "From *PAN* to Pandemic." *Nature,* 397.

Wertheim, J. O., and M. Worobey. 2009. "Dating the Age of the SIV Lineages That Gave Rise to HIV-1 and HIV-2." *PloS Computational Biology,* 5 (5).

Willrich, M. 2011. *Pox: An American History.* New York: The Penguin Press.

Wolf, R. H., B. J. Gormus, L. N. Martin, G. B. Baskin, G. P. Walsh, W. M. Meyers, and C. H. Binford. 1985. "Experimental Leprosy in Three Species of Monkeys." *Science,* 227.

Wolfe, N. 2011. *The Viral Storm: The Dawn of a New Pandemic Age.* New York: Times Books/Henry Holt.

Wolfe, N. D., C. Panosian Dunavan, and J. Diamond. 2004. "Origins of Major Human Infectious Diseases." *Nature,* 447.

Wolfe, N. D., W. M. Switzer, J. K. Carr, V. B. Bhullar, V. Shanmugam, U. Tamoufe, A. Tassy Prosser, et al. 2004. "Naturally Acquired Simian Retrovirus Infections in Central African Hunters." *The Lancet,* 363 (9413).

Woolhouse, M. E. J. 2002. "Population Biology of Emerging and Re-emerging Pathogens. *Trends in Microbiology,* 10 (10, Suppl.)

Worobey, M. 2008. "The Origins and Diversification of HIV." In *Global HIV/AIDS Medicine,* ed. P. A. Volberding, M. A. Sande, J. Lange, W. C. Greene, and J. E. Gallant. Philadelphia: Saunders Elsevier.

Worobey, M., M. Gemmel, D. E. Teuwen, T. Haselkorn, K. Kuntsman, M. Bunce, J.-J. Muyembe, et al. 2008. "Direct Evidence of Extensive Diversity of HIV-1 in Kinshasa by 1960." *Nature,* 455.

Wrong, M. 2001. *In the Footsteps of Mr. Kurtz: Living on the Brink of Disaster in Mobutu's Congo.* New York: HarperCollins.

Zhu, T., and D. D. Ho. 1995. "Was HIV Present in 1959?" *Nature,* 374.

Zhu, T., B. T. Korber, A. J. Nahmias, E. Hooper, P. M. Sharp, and D. D. Ho. 1998. "An African HIV-1 Sequence from 1959 and Implications for the Origin of the Epidemic." *Nature,* 391.

ACKNOWLEDGMENTS

Although *The Chimp and the River* focuses primarily on the ecological origins of HIV, and on the scientific work done to trace those origins, the scope of the human catastrophe that is the AIDS pandemic must be noted again, first and foremost, even here in a brief note to record literary debts. We all stand chastened, grieved, and diminished by the miseries and losses that have been suffered by our fellow men and women, as well as awed by and grateful for the courage, determination, and heart of those who have fought against this catastrophe in so many ways.

As noted above, a number of busy scientists gave their generous cooperation to my research toward this book, by sitting for interviews or responding to e-mail or telephone questions: Robert Gallo, Jane Goodall, Beatrice Hahn, Jean-Marie Kabongo, Phyllis Kanki, Brandon Keele, Elizabeth Lonsdorf, Martin Muller, J. J. Muyembe, Martine Peeters, Jane Raphael, Dirk Teuwen, Karen Terio, and Michael Worobey. Most of those people, in addition, did me the vital favor of reading and correcting draft pages. Three other scientists, whose work does not directly involve AIDS, also read this book in its original form (as a long chapter of *Spillover*) and offered keen editorial advice: Charlie Calisher, Mike Gilpin, and Jens Kuhn. I'm deeply grateful to them all.

My gratitude extends also to Patrick Atimnedi, Anton Collins, Zacharie Dongmo, Ofir Drori, Mike Fay, Barbara Fruth, Shadrack Kamenya, Iddi Lipende, Julius Lutwama, Pegue Manga, Neville Mbah, Apollonaire Mbala, Achile Mengamenya, Jean Vivien Mombouli, Albert Munga, Max Mviri, Hanson Njiforti, Moïse Tchuialeu, and Lee White, all of whom assisted my inquiries about HIV's origins in Africa. There were others who helped in many ways during the broader effort of researching *Spillover* (including my editors and other colleagues at *National Geographic*, among whom that larger project had its beginning), and though their fields of expertise or activity lay outside the immediate frame of the HIV/AIDS story, they contributed much toward allowing me to place that story within its appropriate context: as the most consequential of all modern instances of zoonotic disease. I thanked them by name in *Spillover*, and I thank them again collectively here.

My other signal debts of gratitude are to Maria Guarnaschelli, my longtime editor at W. W. Norton, who gave her keen eye, her astute judgment, and her literary passion to this book as well as a half dozen others we've collaborated on over the past twenty-five years; and to Amanda Urban, my wonderfully ferocious and smart agent at ICM. Many other people at Norton and ICM have also contributed to this project, and I very much appreciate their work. Renée Golden, Binky Urban's predecessor in my professional life, helped me for decades along the route that led toward this sort of project. Gloria Thiede, faithful Gloria, transcribed all the interviews quoted here. Emily Krieger combined assiduous research with a reader's sense of flow, both crucial, in serving as my fact-checker. Daphne Gillam drew the artful, human-handed map. My amazing wife Betsy was always nearby to listen, to read, to discuss, to counsel, and to hold the family together, even while fulfilling her own professional duties. Harry, Nick, Stella, Oscar: Thanks for all you give. I've been blessed with an extraordinary network of colleagues, sources, professional partners, friends, and loved ones, and I'm quite aware that the work couldn't happen without them.

INDEX

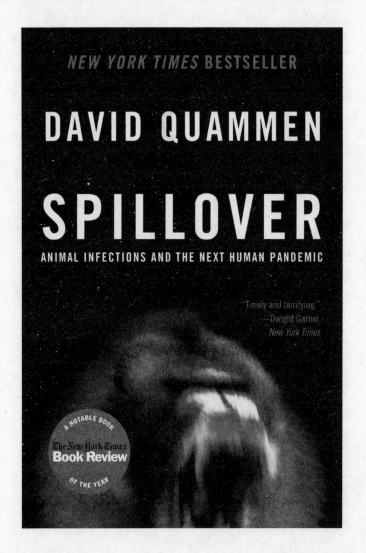